The Water You Drink

The Water You Drink

SAFE OR SUSPECT?

Julie Stauffer

NEW SOCIETY PUBLISHERS

Cataloguing in Publication Data:
A catalog record for this publication is available from the National Library of Canada.

Copyright © 2004 by Julie Stauffer.
The idea for *The Water You Drink* originated with the Centre for Alternative Technology in Wales, UK, where it was published in a significantly different form under the title *Safe To Drink?*
All rights reserved.

Cover design by Diane McIntosh. Image ©Digital Vision

Printed in Canada.

New Society Publishers acknowledges the support of the Government of Canada through the Book Publishing Industry Development Program (BPIDP) for our publishing activities.

Paperback ISBN: 0-86571-497-5

Inquiries regarding requests to reprint all or part of *The Water You Drink* should be addressed to New Society Publishers at the address below.

To order directly from the publishers, please add $4.50 shipping to the price of the first copy, and $1.00 for each additional copy (plus GST in Canada). Send check or money order to:
New Society Publishers
P.O. Box 189, Gabriola Island, BC V0R 1X0, Canada
1-800-567-6772

DISCLAIMER
Although every effort has been made to verify the accuracy of the material presented in this book, neither the author nor the publishers assume responsibility or liability for any damages arising from the use of this material.

New Society Publishers' mission is to publish books that contribute in fundamental ways to building an ecologically sustainable and just society, and to do so with the least possible impact on the environment, in a manner that models this vision. We are committed to doing this not just through education, but through action. We are acting on our commitment to the world's remaining ancient forests by phasing out our paper supply from ancient forests worldwide. This book is one step towards ending global deforestation and climate change. It is printed on acid-free paper that is 100% old growth forest-free (100% post-consumer recycled), processed chlorine free, and printed with vegetable based, low VOC inks. For further information, or to browse our full list of books and purchase securely, visit our website at: www.newsociety.com

NEW SOCIETY PUBLISHERS   www.newsociety.com

Contents

ACKNOWLEDGMENTS viii
INTRODUCTION 1
CHAPTER ONE: LIFE BLOOD 3
    WHAT IS WATER? 3
    WHERE DOES IT COME FROM? 3
CHAPTER TWO: WHAT'S IN IT? 7
    DRINKING WATER CHARACTERISTICS 7
        Dissolved Minerals 7
        Dissolved Gases 8
        Suspended Particles 8
        Color 8
        Taste and Smell 9
        Hardness 9
        pH 10
    DRINKING WATER CONTAMINATION 10
        Lead 13
        Aluminum 16
        Nitrates 17
        Trihalomethanes 19
        Pesticides 20
        Pathogens 22
        Cryptosporidium 24
        Giardia 25
        E. coli 26
        Arsenic 27
        Radon 28
        Perchlorate 29
CHAPTER THREE: HOW IS IT TREATED? 31
    THE NATURAL CLEANSING PROCESSES 31
    WATER TREATMENT PLANTS 32
        Preliminary Screening and Straining 32
        Coagulation and Flocculation 33
        Sedimentation or Flotation 33

  Filtration 33
  Disinfection 35
  Additional Treatment Processes 39
  Fluoridation 39
 SPECIAL CHALLENGES FOR SMALL SYSTEMS 41
 THE MULTIBARRIER APPROACH 41

CHAPTER FOUR: THE DISTRIBUTION SYSTEM 43
 CORROSION 44
 CROSS-CONNECTIONS AND BACKFLOW 44
 BREAKS AND LEAKS 46
 BIOFILM 46

CHAPTER FIVE: PROTECTING THE SOURCE 47
 PREVENTION PAYS 47
 COMMUNITY SOURCE PROTECTION 48
 STATE AND PROVINCIAL LEGISLATION 50

CHAPTER SIX: TESTING DRINKING WATER 53
 WHICH CONTAMINANTS? 53
 HOW OFTEN? 54
 BY WHOM? 55
 WHO SHOULD KNOW THE RESULTS? 55

CHAPTER SEVEN: HOW IS IT MANAGED? 59
 REGULATING WATER 59
 SETTING STANDARDS 60
 PAYING FOR WATER 61
  Who Pays What? 62
  Water Metering 62
 PRIVATIZATION 64
  Private Construction, Operation and Maintenance 65
  Private Operation and Maintenance 66
  Private Financing 66
 BULK WATER EXPORTS 69

CHAPTER EIGHT: THE BOTTOM LINE 71
 ARE THE REGULATIONS STRICT ENOUGH? 71
 ARE WATER SUPPLIERS MEETING REGULATIONS? 73
 WHAT IF YOU'RE PARTICULARLY VULNERABLE? 74

CHAPTER NINE: MAKING DO WITH LESS 77
IS THERE ENOUGH? 77
Water Consumption 77
Up Against the Limit? 79
REDUCING WATER CONSUMPTION IN THE HOME 80
In the Bathroom 80
In the Laundry Room 84
In the Kitchen 84
In General 84
Outside 85
REUSING WASTEWATER 86

CHAPTER TEN: IF YOU'RE NOT HAPPY 89
FINDING OUT WHAT'S IN IT 89
MAKING A COMPLAINT 91
GETTING INVOLVED 91

CHAPTER ELEVEN: ALTERNATIVES TO TAP WATER 93
BOTTLED WATER 93
FILTER SYSTEMS 98

CHAPTER TWELVE: IF YOU'RE NOT ON THE MAINS 101
OBTAINING AN INDEPENDENT SUPPLY 101
Groundwater 102
Surface Water 108
Rainwater 110
STORING AND DISTRIBUTING YOUR SUPPLY 114
MAKING IT DRINKABLE 115
Point-of-Entry Treatment 116
Point-of-Use Treatment 117
TESTING YOUR WATER SUPPLY 122

CONCLUSION 125

GLOSSARY 129

RESOURCES 131

APPENDIX A:
US and Canadian Standards for Drinking Water 136

INDEX 145

Acknowledgments

THE WATER YOU DRINK is based on a similar book I wrote for UK consumers that was published in 1996 by the Centre for Alternative Technology, although this new version is considerably revised and updated for a North American audience. I'd like to thank Dave Thorpe, the editor of that original book who first proposed the idea and shepherded it through to publication, and Graham Preston, who created the illustrations. I'm also grateful to Chris Weedon and Clive Newman, who both reviewed the UK manuscript.

This current manuscript was kindly reviewed by Randy Christensen of the Sierra Legal Defence Fund and Hongde Zhou, a professor of engineering at the University of Guelph. Their comments and suggestions were extremely helpful, and I thank them both for their input.

Finally, I'd like to thank Chris Plant at New Society Publishers for tracking me down and suggesting I write a North American version of *Safe to Drink?* — and then waiting patiently until I had the time to sit down and do it.

Introduction

DRINKING WATER CONCERNS US ALL. We need clean water to live, and we need it every day. However, more and more people are losing faith in the quality of their tap water. In a recent survey, almost one quarter of Americans said they do not drink water straight from the tap because of aesthetic or health concerns. Sales of home filtering systems and bottled water are booming.

Should we be worried about our water supply? On one hand, the answer is yes. Treatment systems do fail from time to time, resulting in occasional outbreaks of cryptosporidiosis or giardiasis, for example. There are also long-term health concerns. Thousands of chemicals are finding their way into our water supplies every year. Some cause cancer, some may disrupt hormone systems, some are toxic in very small quantities. And because many of these chemicals don't break down quickly, they build up in the water cycle. If nothing is done, the situation will continue to get worse.

Nonetheless, there are also plenty of reasons for optimism. Modern treatment systems ensure that the water in our taps, generally speaking, is clean, reliable and plentiful. Waterborne epidemics no longer claim thousands of lives as they once did — you're not going to get cholera or typhoid fever from your drinking water. Drinking water is tested for dozens of pathogens and chemicals, and in the vast majority of cases it comes through with flying colors. Overall, the quality of tap water in North America is very good.

That doesn't mean we can take it for granted. In his inquiry into one of Canada's worst outbreaks of waterborne disease — an *E. coli* outbreak in Walkerton, Ontario — Judge Dennis O'Connor writes: "The Walkerton experience warns us that we may have become victims of our own success, taking for granted our drinking water's safety. The keynote in the future should be vigilance. We should never be complacent about drinking water safety."

This book gives you the facts about your drinking water. It describes how water gets from its source to your tap. It explains how drinking water can become contaminated, what the most common contaminants are, and which ones are worth worrying about. It looks at the management and regulation of the water industry. It tells you what to do if you're not happy with your supply, and it weighs the pros and cons of bottled water and filtered water.

Finally, it outlines the process of obtaining, storing and treating an independent water supply if you're not on the mains. You'll find a glossary of technical terms at the back of the book, along with some helpful resource listings.

CHAPTER ONE

Life Blood

WHAT IS WATER?

WATER — TWO HYDROGEN ATOMS combined with an oxygen atom — is an amazingly versatile, surprising and powerful substance.

Life on earth would be impossible without it. It forms 70 percent of the mass of the human body, for example, carrying salts, gases, minerals and many other substances around our bodies. It's also an essential part of the environment around us: nourishing plants and animals, providing a home to fish and other aquatic life, and playing a key role in the Earth's natural cycles. Since humans haven't yet found a substitute for water, we've all got very compelling reasons for wanting safe and plentiful supplies of water.

WHERE DOES IT COME FROM?

Water seems like a plentiful resource — after all, much of the Earth's surface is covered in it. But although there's lots of water around, most of it is not suitable for drinking (see Figure 1.1). Ninety-seven percent is salt water. Of the three percent that is freshwater, most of that is frozen, leaving less than one percent available for drinking purposes. And of that, only a fraction is easily accessible. So potential sources of drinking water are actually quite limited. They are also shrinking as more and more sources become too polluted to be used for drinking water, and as salt water contaminates freshwater supplies.

North America is often thought to be a region rich in water resources. However, according to the United Nations, 40 percent of North Americans are living in areas with severe water stress, which it defines as areas where more than 40 percent of the water in a river basin is being removed for human use. We may have water resources, but it seems we're not managing them very wisely. Climate change will only make this situation worse in the middle of the continent, where it's going to mean less rainfall and higher temperatures in the summer.

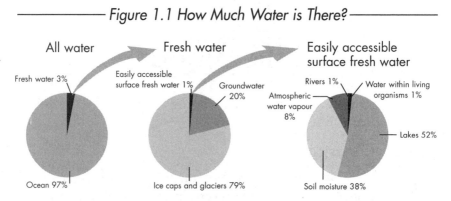

Figure 1.1 How Much Water is There?

Accessible freshwater is only a tiny fraction of the total global water resources.

Not only is our supply of drinkable water limited, it's also recycled. New water is not being created at a significant rate, nor is existing water destroyed for ever. We just use the same water over and over as it travels through the water cycle. For this reason it is quite likely that at some point in your life you may have drunk molecules of what was once Abraham Lincoln's bathwater.

There are three major reservoirs of water — atmospheric, land and oceanic — and water constantly cycles from one to another (see Figure 1.2, below). When water falls to earth as rain, approximately 10 percent filters through the ground, 20 percent runs into lakes, rivers and oceans, and 70 percent returns to the atmosphere through evaporation and transpiration. But the water in the ground eventually filters into the lakes and rivers, lakes and rivers flow to the ocean, and water in lakes, rivers and oceans evaporates into the atmosphere, where it condenses to form clouds. The cycle is complete when the water falls back to earth as rain.

Figure 1.2 The Hydrological Cycle

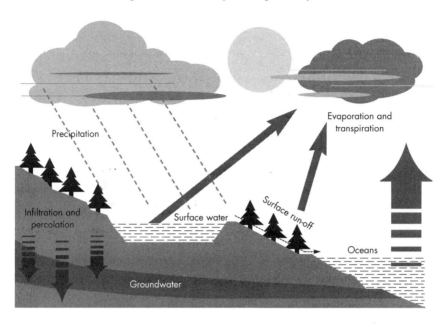

Water constantly cycles between clouds, surface water, underground aquifers, and oceans. This hydrological cycle is driven by the energy of the sun.

Drinking water comes mainly from two sources: surface water and groundwater. Surface water consists of lakes, reservoirs, rivers and streams. About two-thirds of Americans get their drinking water from surface water; in Canada the number is closer to four-fifths. The rest rely on groundwater — water trapped beneath the Earth's surface in porous, water-bearing rocks called aquifers, accessed through wells and boreholes.

As we'll see in the following chapters, different water sources have different characteristics — they can be clean or contaminated, pleasant or unpleasant to taste, clear or silty, hard or soft. A good water treatment system needs to take all this into account in order to produce safe, drinkable water.

CHAPTER TWO

What's In It?

CHEMICALLY SPEAKING, water consists of two molecules of hydrogen and one molecule of oxygen. Pure, distilled water is just that — hydrogen and oxygen. It's odorless and tasteless and not very pleasant to drink. But the water you find in nature is not pure H_2O — there's lots of other things in there too. Some are good, some are bad. Some are natural, and some we have to blame on ourselves. In this chapter, we'll look at the most important stuff that can show up in our water supplies.

DRINKING WATER CHARACTERISTICS

Let's begin by focusing on the natural characteristics. Take a glass of water from your local river, lake or well. It can contain dissolved minerals and gases and suspended particles. The water may be hard or soft, acidic or alkaline, clear or cloudy. It may contain lots of organic material or none at all, and there may or may not be micro-organisms present.

All these characteristics determine the overall quality of the water: how good it tastes, how clean it looks, and how safe it is to drink. These characteristics vary from source to source and from day to day, and they will help to determine what kind of treatment will be required to make that water drinkable.

Dissolved Minerals

There are lots of minerals in soil that can dissolve in the water as it passes through, contributing to its taste. Some mineral-rich waters are bottled and sold as "mineral water" precisely because they taste good.

The most common minerals found in water are iron, manganese, silica, nitrate and fluoride, as well as ions of calcium, magnesium, sodium, bicarbonate, sulfate and various trace elements. Some of these elements are necessary for health (although we get most of our mineral requirements from food). Some create problems — too much iron or manganese, for example, will stain laundry and porcelain fixtures brown. Others, such as lead, mercury and cadmium, are toxic. Total dissolved solids in natural waters can range from 10 mg/L to over 100,000 mg/L, but if the concentration is higher than 500 mg/L, the water often has an unpleasantly strong taste.

Dissolved Gases

When surface water mixes with air, gases become dissolved in the water. The most common dissolved gases are oxygen, nitrogen and carbon dioxide, because they are most abundant in the air. They add to the quality of drinking water — without dissolved gases, water tastes "flat." Some groundwater supplies have enough dissolved carbon dioxide that the water is actually fizzy.

Suspended Particles

Suspended particles make the water appear cloudy or muddy. Technically, this is called turbidity. It can be caused by inorganic material such as mineral sediments or iron oxides, or by organic material such as algae. Suspended particles are taken out during water treatment, but treatment plants often struggle to keep turbidity levels low after heavy rainfalls or during spring snowmelt, when lots of soil particles wash into surface water.

You may notice a stronger chlorine taste in your water in the spring, when treatment plants often add extra disinfectant to kill the micro-organisms that are carried along with the soil particles.

Color

Color is usually due to one of two things: organic material leaching from peat and other decaying vegetation; or metallic salts of iron and manganese. Leaves tend to give water a reddish color and iron, not surprisingly, gives it a rusty hue. Manganese can combine with chlorine to turn your water black. All this stuff is relatively harmless, but it can stain clothing and porcelain. Color often changes during the year — in autumn, for example, water tends to be more colored because surface waters are filled with decaying leaves.

Taste and Smell

Many things can give water unpleasant tastes and odors, although it can still be perfectly safe to drink. Algae and molds create swampy or musty tastes; iron, manganese and sulfates cause bitter tastes; hydrogen sulfide (produced by certain kinds of bacteria) has an unmistakable rotten-egg smell; and chlorine residuals give water a chlorine taste, like water in a public swimming pool. Water treatment plants should remove most objectionable tastes and odors, but they are sometimes responsible for adding chlorine tastes, particularly if the chlorine reacts with other chemicals in the water.

Hardness

Dissolved calcium and magnesium salts, as well as strontium, iron and manganese cause hardness (see Figure 2.1, below). Hard water is good to drink, but too much hardness causes a couple of problems. One is scum: calcium can combine with fatty acids in soap to form scum on sinks, bathtubs and clothing, making washing difficult. The other is scale: both calcium and magnesium can combine with bicarbonates, carbonates, sulfate and silica in the water to form hard deposits in pipes and kettles. (See *Making It Drinkable,* Chapter Twelve, for a discussion of water softeners.)

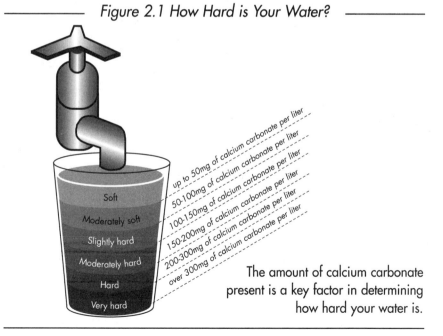

Figure 2.1 How Hard is Your Water?

The amount of calcium carbonate present is a key factor in determining how hard your water is.

In contrast, softness is caused by low levels of calcium and magnesium. Soft water is better for washing and it suds very well, but it is not so healthy to drink. It is linked to heart disease, and it can corrode lead pipes, increasing the level of lead in your drinking water. So water suppliers have to strike a balance and provide water soft enough to use for washing but hard enough to be healthy.

pH

pH describes how acidic or basic (see Glossary) a sample of water is, measured on a scale from one to fourteen. Water with a pH lower than seven is defined as acidic; water with a pH higher than seven is basic. Ideally, drinking water should be slightly basic, and both Canada and the US recommend treatment plants keep pH between 6.5 and 8.5.

Acidity is usually the result either of carbon dioxide dissolving to produce weak carbonic acid, or of organic acids produced when leaves and other vegetation decompose. Water with a pH less than 6.5 can cause corrosion in metal pipes, increasing the metal content of your drinking water. Surface waters from peaty moorlands may have a pH as low as four.

Water with a pH above 8.5 will have a strong caustic taste and cause scale formation in pipes. Hard groundwater that has percolated through chalk or limestone can have a pH as high as nine.

DRINKING WATER CONTAMINATION

Certain natural characteristics contribute to the quality of water, making it good to drink, or bad to drink. But there are other things that get into water that also affect its quality. These are contaminants — any undesirable physical, chemical or microbiological substance. All kinds of contaminants can show up in a drinking water source: industrial chemicals, animal sewage ... even traces of pharmaceutical drugs. We're not talking about large quantities — usually only a few thousandths of a gram in a liter of water — but in the case of certain contaminants, that's enough to be dangerous.

Contaminents make their way into water sources in four main ways (see Figure 2.2). The most obvious source of contaminants is effluent — the wastewater discharged by industry and sewage treatment plants into local rivers or lakes. It may be quite clean or quite polluted, depending on how much it's treated before it goes out the pipe. Sewage effluent is a major polluter, but industrial and agricultural effluent and acid drainage from coal and metal mines also contribute to poor water quality.

Figure 2.2 Measuring Contaminants

Contaminants get into water sources from:

EFFLUENT   LEACHING   RAINWATER   RUNOFF

Contaminants are usually present in extremely small quantities — so small that until recently they couldn't be measured. Most contaminants are measured in units of milligrams (one thousandth of a gram) per liter of water (mg/L) or micrograms (one millionth of a gram) per liter of water (μg/L). One milligram of contaminant per liter means one part per million. One microgram per liter is even smaller: one part per billion.

Another common culprit is runoff — water that travels over the surface of the land after a rainfall or during the spring snowmelt, picking up contaminants along the way. Urban runoff can contain lots of undesirable things: dirt and grit, dog feces, disease-causing micro-organisms, heavy metals, pesticides, rubber particles from automobile tires, and gasoline and oil from leaking tanks. Agricultural runoff also causes significant pollution because it can contain large quantities of pesticides, fertilizers and animal wastes.

In the case of groundwater contamination, leaching is to blame. When water percolates through a material — such as the soil — it dissolves various substances in the process. Once these contaminants leach into groundwater it is very difficult to undo the damage, partly because the micro-organisms that help to cleanse surface water do not live underground. Common sources of groundwater pollution include leaking from cesspools; leachate from waste dumps, sewage lagoons, industrial lagoons and landfills; as well as accidental spills, leaking underground storage tanks, mining, agriculture, and the movement of polluted surface water into underground supplies.

Finally, precipitation can be another source of pollution in areas with poor air quality. When rain or snow falls through polluted air, it can pick up a whole host of contaminants including pesticides, asbestos dust, lead, chlorinated hydrocarbons, carbon monoxide, sulfur dioxide (the major cause of acid rain), nitrous oxides and radioactive fallout.

Since water is constantly recycled between the atmosphere, the soils and the oceans, pollution that contaminates one part of the water cycle can eventually affect the entire system. Natural purification processes, sewage treatment plants and water treatment plants help to cleanse water, but they can't take out everything. So what we dump down our drains can ultimately come back to us through our taps.

In the nineteenth century, the biggest threats to drinking water supplies were from bacteria and viruses that caused disease. Typhoid and cholera were major killers, carried by drinking water contaminated with untreated sewage. Today, modern water treatment methods have eliminated all but a few waterborne diseases in developed countries, and now the focus of concern has shifted to chlorine-resistant micro-organisms like *Cryptosporidium*, *Giardia* and viruses, and to the possible long-term effects of small quantities of chemical contaminants.

The following is a list of common contaminants in your drinking water that may be of concern. These are only a handful of the contaminants that can be present in your drinking water, but they are some of the more important ones.

You'll find a description of where they come from, what they're likely to do to you, how they can be removed, and what American regulations and Canadian guidelines say are acceptable levels. Where applicable, provincial and territorial standards are also included. Note that in Canada, drinking water is regulated at the provincial level. The federal government has created a set of guidelines that designate Maximum Acceptable Concentrations (MACs) for several dozen contaminants, but it is up to each province which guidelines, if any, to adopt (see *Setting Standards,* Chapter Seven). The result is a wide variation across the country. If a province chooses adopt

---

Care with Household Hazardous Waste

Avoid putting chemicals down the drain! What you put down there can harm the aquatic environment, it can interfere with sewage treatment, and it can turn up in your drinking water. Motor oil, paints and solvents in particular should never go down the drain. Many municipalities have special facilities to deal with these types of hazardous household waste. Old or unused drugs should be returned to your pharmacist, not flushed down the toilet. Trace amounts of prescription drugs, including anticonvulsants and cholesterol medication, have shown up in water samples from ten Canadian communities. Even many cleaning products are harmful, including floor cleaners, oven cleaners and bleach. Make sure you use only safe, biodegradable alternatives – read the labels!

MACs as regulations or standards, they become legally enforceable. If instead, they adopt MACs as guidelines or objectives, they have no legal power. In contrast, drinking water is regulated federally in the United States, and the Maximum Contaminant Levels (MCLs) are legally binding.

It's worth mentioning that the jury is still out on the health effects of many of these substances. One study may show a substance is linked to cancer in humans, another may not. The problem is finding good evidence (see *Deciding Which Chemicals Cause Cancer*, page 14). It's also important to mention that children tend to be more susceptible to dangerous contaminants than adults because their immune systems are not as strong and they consume more water, per kilogram of body weight, than adults.

Lead

HEALTH EFFECTS

Lead is probably the most dangerous of the metal contaminants in North American drinking water. It is a poison that accumulates in the body and affects the nervous system, causing mental retardation and behavioral problems in young children, and increasing the risk of still births and low-birthweight babies. It can also cause anemia, hypertension, and kidney dysfunction.

SOURCE

Drinking water accounts for approximately 20 percent of our exposure to lead in North America, and most lead contamination in drinking water is due to lead in household pipes. The use of lead in pipes and solder has been banned in both the US and Canada, but it's still present in the plumbing in many older houses. It is also present in solder used to join copper pipes. If your water is soft and acidic, lead will leach out of household pipes more readily, but this can also occur in hard water areas. Lead pipes may also be present in your water supplier's distribution system — according to the US Environmental Protection Agency, approximately 20 percent of all public water distribution systems in the United States contain some lead components.

If you are concerned, check your household plumbing (see *Worried About Lead?* page 15), and contact your water supplier for more specific information.

Deciding Which Chemicals Cause Cancer

When the World Health Organization (WHO) tries to decide which chemicals cause cancer in humans, it looks at two kinds of evidence: studies of cancer in human populations ("epidemiological evidence") and experimental research on laboratory animals. Both types of evidence have their weaknesses. When it comes to epidemiological studies, it's easy to say that a particular group of people has a higher risk of cancer, but it's difficult to say what causes that risk, especially because cancers take many years to develop. When it comes to animal studies, the problem is that it's not always possible to say that something that causes cancer in guinea pigs when it's fed to them in large doses for a short period of time will cause cancer in humans when it's consumed in small doses over long periods of time. The WHO weighs all the evidence and then classifies chemicals into four groups and two subgroups, according to the following criteria:

- **Group 1** *chemicals cause cancer. Something is classified as Group 1 if there is sufficient evidence that it causes cancer in humans.*
- **Group 2** *chemicals may cause cancer.*
- **Group 2A** *chemicals probably cause cancer, based on limited evidence in humans or sufficient evidence in animals.*
- **Group 2B** *chemicals possibly cause cancer, based on limited evidence in humans or less than sufficient evidence in animals.*
- **Group 3** *chemicals are substances that can't be classified because there is not enough evidence in humans or animals.*
- **Group 4** *chemicals probably don't cause cancer. Something is classified as Group 4 if the evidence suggests it doesn't cause cancer in humans or animals.*

ACCEPTABLE LEVELS

Under US EPA regulations, water suppliers must sample tap water regularly from points throughout the distribution system. If the level of lead exceeds 1.5 ppb in more than 10 percent of samples, the supplier must take "corrective action." Generally, this means increasing the pH of the drinking water, since the more acidic the water is (or, the lower the pH level), the more lead it will leach out of your pipes. If this is not sufficient, the water supplier must replace some of its lead distribution pipes. This approach ensures that most homes have safe levels of lead, but individual homes may still have problems. In Canada, the approach is more straightforward: the Maximum Acceptable Concentration (MAC) for lead is 0.01 milligrams per liter.

TREATMENT

If you suspect that there may be lead in your tap water, the first thing to do is check your home for lead pipework (see *Worried About Lead?* right). You may also wish to have your water tested. In the US, water suppliers are obliged to sample your tap water for lead if you request it, although you'll probably be responsible for the costs involved. In Ontario, water suppliers are obliged to replace lead service lines if a consumer's drinking water contains more than 0.01 mg/L of lead after the municipal pipes have been flushed.

Ideally, if a problem exists, you should replace your household plumbing, especially if you

Worried About Lead?

Checking Your Home for Lead Pipework
There are two simple checks to find out if your water plumbing is lead. An unpainted lead pipe is dull grey and soft. If you scrape the surface gently with a knife, you will see that the dull, soft, silver-colored metal beneath becomes shiny when scratched.

1. *Look in or behind cupboards in your kitchen. Locate the pipe leading to the kitchen tap, and check as much of it as possible.*
2. *Locate the stop valve outside your property. If it's possible, examine the piping from the stop valve to your property.*

Pipes may also be:
- *copper – reddish brown and hard.*
- *iron – dark, very hard and possibly rusty.*
- *plastic – grey, black or blue.*
- *galvanized metal – grey and usually fitted together with threaded joints.*

have young children. Unfortunately, the plumbing inside your house is your responsibility, or your landlord's, and replacing it is an expensive solution. If this isn't possible, or it's not an affordable option, consider a home filter system that meets the ANSI/NSF standards for removing lead. At a minimum, you should run the taps for several minutes before using the water for drinking or cooking, particularly in the morning when the water has been sitting in the pipes overnight. (From a water conservation perspective, this solution is far from ideal, but it will protect your health.) Never drink or cook with water from the hot water tap — it will contain much higher levels of metals.

If you have any further doubts or queries concerning lead piping, contact your water supplier or a qualified plumber. In the United

States, you can also contact the National Lead Information Center (see Resources).

Aluminum

HEALTH EFFECTS

Many people worry about aluminum levels in their drinking water because of a possible link between aluminum and Alzheimer's disease. Some epidemiological studies have found that there is a slightly higher rate of dementia in communities with high levels of aluminum in their drinking water (although other studies have not), and autopsies have shown that Alzheimer's patients have high levels of aluminum in their brains.

On the face of it, the amount of aluminum you get from your drinking water is quite small, since 97 percent of the aluminum you absorb comes from your food. Pharmaceuticals are another major source. Most people consume about eight milligrams of aluminum a day, but that can skyrocket to as much as five grams a day if you take aluminum-based antacids or buffered aspirin. However, your body absorbs the aluminum from drinking water better than aluminum in food, so in actual fact it may be a significant source.

SOURCE

Aluminum occurs naturally in water. It's the second most common element in the Earth's rocks, it's present in many soils, and it can migrate into water that comes into contact with those soils. However, the biggest source of aluminum in drinking water is from aluminum sulfate used as a coagulant during the water treatment process to remove color and turbidity. Most of this gets removed further on in the treatment process, but traces can remain.

ACCEPTABLE LEVELS

Health Canada does not currently have any health-based guidelines for aluminum, since it has found no consistent, convincing evidence that aluminum in drinking water causes adverse health effects. It does recommend that water suppliers using aluminum-based coagulants should reduce aluminum levels in the treated water to less than 0.1 mg/L. Likewise, the US EPA does not have any health-based regulations, but it sets a non-enforceable standard of 0.05-0.2 mg/L to avoid taste and odor problems. Aluminum is on the list of the EPA's

Contaminant Candidates List, meaning that it is one of a dozen or so contaminants that the EPA considers priorities for evaluation.

TREATMENT

Aluminum can be removed by sedimentation and filtration (see *Water Treatment Plants,* Chapter Three). Unfortunately, there is no cheap or easy way to remove aluminum at the household level. A reverse osmosis system or a distillation system will do the job, but both are expensive solutions.

Nitrates

Nitrates are molecules that occur naturally. Plants need nitrates to grow, so you'll find them in natural and synthetic fertilizers. But while nitrates are good for plants, they may not be quite as healthy for humans.

HEALTH EFFECTS

Nitrates can cause "blue baby syndrome," a potentially lethal condition where a baby's red blood cells can't carry enough oxygen to the body. The baby becomes starved of oxygen and turns blue. Infants fed on formula made with water from a nitrate-contaminated well are at the greatest risk. Some studies have linked high levels of nitrates with miscarriages and birth defects — specifically neural tube defects. In adults, nitrates may also increase the risk of non-Hodgkin's lymphoma, and some studies have found a connection between nitrites (about one quarter of nitrates are broken down in the body into nitrites) and stomach cancer, although the epidemiological evidence for this is weak.

SOURCE

Nitrates in drinking water come from two primary sources. One is organic waste, particularly human sewage and livestock manure. The second and most important source is chemical fertilizers. Americans use more than 10 million tons of chemical fertilizers each year. That's an enormous volume, and we haven't felt the full impact of it yet. Because the use of these fertilizers has substantially increased in the last two decades, the levels of nitrates in our drinking water will continue to rise as they slowly move down through the soil and into our groundwater supplies.

Shallow wells in agricultural areas are at the greatest risk for nitrate contamination. In rural Ontario, for example, 14 percent of wells contain nitrate levels above the regulations. Similarly, a US Geological Survey found 15 percent of domestic wells in agricultural and urban areas had nitrate levels above the MCL. Nitrate levels are particularly high in the intensive agricultural areas of the American Midwest.

ACCEPTABLE LEVELS

In Canada, the MAC is 45 milligrams per liter, while the US sets a much stricter limit of 10 milligrams per liter. Because nitrate is an acute toxin, the Environmental Protection Agency (EPA) does not allow water suppliers to take averages of nitrate levels; instead, all samples must come in under the limit of 10 milligrams per liter.

Some Canadian provinces have followed the EPA's lead: Quebec and Ontario have set a limit of 10 mg/L of nitrates and nitrites combined. Newfoundland has a standard of 45 milligrams per liter, and the Northwest Territories and Nunavut have regulations of 45 milligrams per liter. The other provinces have unenforceable guidelines or objectives of 45 milligrams per liter, and the Yukon and PEI have nothing in place. The World Health Organization recommends a limit of 50 milligrams per liter for nitrates.

If your drinking water contains less than 10 milligrams per liter, vegetables will probably be your major source of nitrate. On the other hand, if the nitrate level in your water is 50 milligrams per liter, you'll get most of your nitrate from drinking water.

TREATMENT

There is no cheap or simple way for water suppliers to reduce nitrate concentrations, except by mixing contaminated water with clean supplies. Specialized treatment such as ion exchange, electrodialysis, and reverse osmosis can be used, but these are much more costly options.

There are steps you can take at home, however. If your tap water contains high levels of nitrates, consider a reverse osmosis system that meets ANSI/NSF standards for nitrate removal, or use bottled water. Don't use nitrate-contaminated water to mix infant formula. Instead, use filtered or bottled water, or buy a ready-mixed formula. Boiling water won't remove nitrates; on the contrary, it will actually increase their concentration. If you rely on a private well for your drinking water, make sure you have it tested regularly for nitrates, particularly in the spring when farmers are using lots of fertilizers.

Trihalomethanes

Trihalomethanes (THMs) are a group of organic compounds containing one carbon atom, one hydrogen atom and three halogen atoms; they include chloroform trichloromethane, (CHC_{l3}), bromodichloromethane ($CHC_{l2}Br$), dibromochloromethane ($CHClBr_2$), and tribromomethane ($CHBr_3$).

Health Effects

According to the World Health Organization, chloroform and bromodichloromethane are "possible" carcinogens. It says if you drink one liter a day at 30μg/L you have a one in 100,000 risk of developing cancer. Health Canada reports that THMs are linked to increased risk of bladder or colon cancer, and the EPA has classified some THMs as probable human carcinogens. Other studies have linked them to miscarriages, birth defects, and low birth weights.

Source

As leaves and other vegetation in surface water break down, they release certain organic substances. THMs are formed when chlorine reacts with these substances. This means that if chlorine is used to disinfect water that contains high levels of organic substances, THMs may be formed. The longer chlorine is in contact with organic materials, the more THMs are formed. Because chlorine is the most common disinfectant used in water treatment, THMs are a widespread concern in drinking water. The danger increases for households further away from treatment plants, since their drinking water stays in contact with chlorine longer as it travels through the distribution system. THM levels tend to be higher in the summer and fall, when there is more organic material in most surface water. However, if you rely on groundwater for your source of drinking water, THMs shouldn't be a big concern, since groundwater contains much lower levels of organic material.

Acceptable Levels

The EPA sets a total MAC of 0.08 milligrams per liter for all THMs combined; in Canada, the interim guideline for total THMs is 0.1 milligrams per liter. Quebec has chosen the stricter US standard of 0.08 milligrams per liter, while the other provinces have gone with 0.1 milligrams per liter. The territories and Prince Edward Island have not set limits for THMs.

Treatment

There are two basic approaches to solving this problem: preventing THMs from forming in the first place, or taking them out once they've been formed. In areas at risk, water suppliers can prevent THMs from forming by settling out, coagulating or filtering the organic material in the water before they disinfect it, so there are fewer organic precursors to form THMs when they add chlorine. Alternatively, they can use non-chlorine disinfectants such as ultraviolet radiation and ozonation so there isn't any chlorine to react with organic material. The other approach is to remove THMs after they are formed — for example, by using activated carbon and air stripping. However, it is important not to compromise disinfection in lowering the level of THMs — the health risk of not disinfecting properly is much greater than the health risk of consuming THMs.

At home, you can reduce the levels of THMs in your tap water with a filter system certified to meet ANSI/NSF standards for total THM (TTHM) reduction. Alternatively, bottled water that comes from an underground source should be free of THMs.

Pesticides

Pesticides are chemicals that kill pests. There are a lot of so-called pests out there, from dandelions to mosquitoes to zebra mussels, and consequently we're spraying large quantities of pesticides into the environment. These include herbicides, which kill weeds; insecticides, which kill insects; and fungicides, which kill molds and fungi. They are widely used on farmland, roadside verges, parks, golf courses and private gardens, and they find their way into water sources in large quantities. Nearly 900 active ingredients are registered in the US, and each year more than half a billion kilograms of pesticides are used in the United States on farms, in businesses, and at home.

Health Effects

Since there are so many different types of pesticides, there isn't room to list all their various health effects here. In fact, the data may not exist, particularly on the long-term effects of drinking low levels. We do know that many are possible carcinogens. Atrazine, for example, may increase the risk of ovarian cancer and disrupt endocrine systems, and may be linked to breast cancer (see *Hormones and Hormone Mimics*, page 23). It is an extremely common pesticide in North America, used to protect corn and soybeans from weeds. Agriculture Canada considers atrazine the greatest pesticide threat to groundwater. 2,4D can damage livers, kidneys and nervous systems. Along with dicamba, 2,4D is the pesticide most commonly found in US public drinking water supplies.

SOURCE

Chemicals have been used as a significant form of pest control since 1945. Today there is almost

Common Pesticides	
Classification	Example
phenoxyacetic acid herbicides	2,4D
	dicamba
triazine herbicides	atrazine
	simazine
organochlorine insecticides	DDT
	aldrin
	dieldrin
	lindane
organophosphate insecticides	malathion

no agricultural area in the world that does not use pesticides, and as a result, pesticides have made their way into many water sources. The US Geological Survey found pesticides in more than 95 percent of stream samples across the country and in almost 50 percent of groundwater samples. Even aldrin, dieldrin and lindane, which are no longer produced in the US, still show up in American drinking water.

ACCEPTABLE LEVELS

Health Canada has not set a guideline for total pesticides, but it has set MACs for more than a dozen individual pesticides. For example, the interim MAC for atrazine and its metabolites is 0.005 mg/L; the interim MAC for 2,4D is 0.1 mg/L; for dicamba the MAC is 0.12 mg/L; and the combined MAC for aldrin and dieldrin is 0.0007 mg/L.

There isn't room here to examine how each of the provinces approach all the different pesticides, but let's look at atrazine as an example. Quebec has set a regulation of 0.005 mg/L, and Ontario likewise has a legally enforceable standard of 0.005 mg/L. Most of the remaining provinces have an unenforceable guideline of 0.005 mg/L, but neither Prince Edward Island nor the territories has any limit set.

The US has regulated approximately two dozen individual pesticides. The limit for atrazine is 0.003 mg/L, for 2,4D it's 0.07 mg/L, and for simazine it's 0.004 mg/L.

The amount of pesticide you are exposed to annually in your drinking water is generally not a health risk, but there can be dangerous seasonal peaks when pesticide applications run off into local watercourses or seep into groundwater. The US Geological Survey found that peak seasonal concentrations in surface waters in agricultural areas frequently exceeded federal drinking water standards.

TREATMENT

Water suppliers can reduce pesticide levels with activated carbon adsorption (see Glossary) or ozonation (see *Water Treatment Plants,* Chapter Three), but some still turn up in drinking water. You can remove pesticides from your tap water by using a jug filter or an in-line filter with activated carbon (see *Filter Systems,* Chapter Eleven and *Making It Drinkable,* Chapter Twelve). Choose a filter that meets the ANSI/NSF standard for the particular pesticides you're concerned about, and maintain it according to the manufacturer's instructions. If you rely on a private well for your drinking water, it might be wise to have the water tested for a few of the most common pesticides, especially if you live in an agricultural area.

Pathogens

Pathogens are any viruses, bacteria, protozoa or other micro-organisms that can cause disease.

HEALTH EFFECTS

There are a whole host of diseases caused by waterborne pathogens. These include cholera, typhoid fever, bacillary dysentery, amebic dysentery, cryptosporidiosis, giardiasis, poliomyelitis and infectious hepatitis. Some particularly worrisome pathogens are described separately below. Although cholera, typhoid and polio are no longer public health threats in North America, waterborne diseases still have a real impact: a recent study in British Columbia found that variations in drinking water quality accounted for 17,500 visits to doctors, 85 admissions to hospital, and 138 trips to pediatric hospital emergency rooms over a six-year period. In the United States, the number of reported waterborne disease outbreaks has increased in recent years: in 1999—2000 there were twice as many as in 1997—1998. This is partially because water suppliers are monitoring their water more vigilantly and are using more advanced detection techniques.

SOURCE

Pathogens can be transmitted by human or animal feces. Whenever sewage or manure contaminates a water supply, there is a danger that pathogens could be present.

ACCEPTABLE LEVELS

Because it would be too difficult and too expensive to test for all possible pathogens, water suppliers test for coliform bacteria instead, because they are naturally present in human and animal feces. So if there are coliforms in the water, it indicates the water has been contaminated with sewage, and other pathogens could therefore be present. According to EPA regulations, total coliforms cannot be present in more than five percent of samples in any given month. In Canada, no more than ten percent of samples in a month can contain total coliforms, and fecal coliforms are not permitted. Most provinces have set similar regulations or guidelines.

TREATMENT

Disinfection will remove most pathogens (see *Water Treatment Plants,* Chapter Three), but not all. *Giardia* and *Cryptosporidium,* for example, are extremely resistant to chlorine disinfection. See specific examples below.

Cryptosporidium

HEALTH EFFECTS

Cryptosporidium is a parasite that causes acute cases of diarrhea, accompanied by headache, abdominal cramps, nausea and vomiting. It is extremely prevalent — surveys have found it in more than 80 percent

---

Hormones and Hormone Mimics

It is now becoming clear that certain chemicals, including many pesticides, can mimic the effects of human hormones. Atrazine, for example, mimics the female sex hormone estrogen. This means that it may interfere with normal human development

High levels of hormone mimics in the environment may also increase cancer rates. For example, most breast cancer is due to high levels of the female hormone estrogen. The more estrogen you are exposed to over your lifetime, the higher the risk that you will develop breast cancer. So exposure to estrogen-mimicking hormones may be contributing to the global increase in breast cancer rates.

Estrogen mimics are also affecting wildlife. A recent American study found a strong link between high levels of atrazine in the Midwest and the unusually frequent appearance of hermaphroditic leopard frogs. In Lake Apopka, Florida, alligators were exposed to a major pesticide spill in 1980. Today, adult alligators produce abnormally low levels of sex hormones — their ovaries and testes appear to be "burnt out."

On top of all the hormone mimics, the prevalent use of birth control pills means our water sources are contaminated with artificial estrogens and progestins, which pass through the human body and are not affected by sewage treatment processes.

of US surface waters tested. In most cases, the levels of *Cryptosporidium* are low enough that they don't cause any problems for healthy individuals, but occasionally, the levels may rise enough to cause an outbreak of cryptosporidiosis.

There is no effective treatment for cryptosporidiosis. It is transmitted by oocysts, a highly resistant form of the parasite. In healthy individuals cryptosporidiosis is unpleasant, but it is not particularly dangerous and will go away without treatment after a week or two. However, it is serious and can be fatal in people with low immunity, such as people with AIDS, chemotherapy patients, the very old and the very young. Anyone with a compromised immune system should check with their physician about the need to boil their water, use special filters or switch to bottled water.

Source

Oocysts are excreted in the feces of infected hosts. Most *Cryptosporidium* oocysts originate from infected livestock.

Acceptable Levels

Canada does not currently have guidelines for acceptable levels of *Cryptosporidium* oocysts, but US regulations require large cities to remove or inactivate 99 percent of the oocysts, and soon all water suppliers will have to meet this standard. Coliforms are not an adequate indicator for the presence of oocysts, because oocysts are far more resistant to disinfection than coliforms. Unfortunately, current methods for detecting *Cryptosporidium* oocysts are not terribly reliable.

Treatment

Water treatment plants can deal with low levels of oocysts, but if the concentration becomes too high, some oocysts will pass into the drinking water supply. Chlorine has no effect, but ozonation and UV irradiation can render the oocysts inactive.

You can remove *Cryptosporidium* from your tap water by boiling it for one full minute (longer at higher altitudes). Alternatively, use a filter system that meets ANSI/NSF standards for cyst removal or that is labelled as "Absolute 1 micron." Reverse osmosis systems are also effective. Don't assume that bottled water is necessarily safe — look for brands that are purified by distillation or reverse osmosis.

Giardia

HEALTH EFFECTS

Giardia is a protozoa that causes diarrhea, abdominal cramps and fever, often referred to as "beaver fever" because beavers are a common source of this parasite. It usually clears up without treatment within a month in otherwise healthy people. For vulnerable patients with suppressed immune systems, antiparasitic drugs can be used to treat the disease.

Case Study: Cryptosporidiosis Outbreak

One of the most serious outbreaks of cryptosporidiosis in the US occurred in Milwaukee in 1993. Heavy rainfall coincided with mechanical and operational problems at one of the city's water treatment plants. As a result, the treated water leaving the plant had higher levels of turbidity than usual, a situation that can often mask high levels of Cryptosporidium *oocysts.*

Although the source of Cryptosporidium *was never tracked down, the impacts were clear. 400,000 people fell ill, more than 4,000 were hospitalized, and 50 died from the waterborne disease. Most of the deaths occurred in people with AIDS, whose immune systems weren't functioning well.*

The Milwaukee outbreak catalyzed the development of the EPA's Interim Enhanced Surface Water Rule, which requires water treatment plants in large centers to remove 99 percent of Cryptosporidium *oocysts.*

More recently, North Battleford, Saskatchewan was the site of a cryptosporidiosis outbreak that affected roughly 6,000 people in 2001. An independent inquiry concluded that "the city lacked an appreciation that safe drinking water is a public health priority," the water treatment plant operations were substandard, and drinking water was ineffectively regulated by the province. Just over a year after the outbreak, Saskatchewan announced the implementation of a Long Term Safe Drinking Water Strategy that includes stronger regulations governing drinking water quality.

SOURCE

It is found in the feces of infected mammals — beavers, muskrats, dogs and humans — in the form of small, resilient cysts.

ACCEPTABLE LEVELS

US regulations call for a 99.9 percent reduction or inactivation of *Giardia* cysts, but there are no similar guidelines in Canada.

TREATMENT

Water suppliers can remove over 99.9 percent of cysts by slow sand filtration. Coagulation followed by rapid gravity filtration is less effective and does not meet EPA regulations. *Giardia* cysts are more susceptible to disinfection than *Cryptosporidium*. However, as with the tests for *Cryptosporidium,* the tests available for detecting *Giardia* are not reliable. If you're particularly vulnerable to disease, follow the same water treatment guidelines outlined above (see *Cryptosporidium*).

E. coli

HEALTH EFFECTS

There are hundreds of different strains of *E. coli*. While most are harmless, *E. coli* O157:H7 produces a powerful toxin that can be lethal. This strain is best known as a contaminant in undercooked ground beef that causes "hamburger disease," but it crops up occasionally in water sources. It causes diarrhea lasting approximately four days, and often bloody diarrhea and severe abdominal pain, as well. In vulnerable sectors of the population, such as children and the elderly, it can cause anemia, acute kidney failure, and in some cases death.

SOURCE

E. coli O157:H7 is found in the intestines of animals and humans, and therefore can be present in water sources contaminated with sewage or manure.

ACCEPTABLE LEVELS

E. coli O157:H7 is a type of fecal coliform. Canadian guidelines do not permit any fecal coliforms in drinking water. EPA regulations do not permit total coliforms in more than five percent of samples in any given month.

TREATMENT

Chlorination, UV irradiation and ozonation will all kill *E. coli* O147:H7. At home, boiling your water is also effective, but this will increase the concentration of other contaminants. It's fine as a short-term solution, but consider using another form of disinfection over the long term.

> ### Case Study: E. coli Contamination
>
> *In May 2000, seven people died and more than 3,000 became ill in the small town of Walkerton, Ontario as a result of E. coli O157:H7 contaminating their drinking water. As in most cases of waterborne disease, it was the very young and the very old who were most severely affected.*
>
> *The source of contamination was traced to manure spread on a field near one of the wells that provided the town's water supply. The farmer spreading the manure followed proper procedures, but the well was vulnerable to contamination and was not monitored properly. An inquest into the tragedy revealed a water treatment system rife with problems: inadequately trained personnel, poor monitoring practices, inadequate disinfection, and falsification of records. Furthermore, the public utilities manager in charge of the water treatment plant failed to act for several days after he received test results showing contamination, and denied any problem existed, greatly increasing the extent of the outbreak.*
>
> *Other factors contributed as well: the private lab that handled the water testing was not required to advise either the Ministry of Environment or the local medical health officer of the adverse test results, as is the law in many other jurisdictions. Nor was there proper oversight. Although the Ministry of the Environment inspected the water system on several occasions, the inspectors failed to identify problems, or when they did identify problems, they failed to follow up on them.*
>
> *The economic impact of the tragedy was calculated at more than Can$64.5 million (US$47 million), and many of the people who fell ill will have long-term health effects, including lasting damage to their kidneys. Following the Walkerton outbreak, the Ontario government tightened up regulations on testing and treating drinking water.*

Arsenic

HEALTH EFFECTS

Arsenic is widely recognized as a carcinogen, increasing the risk of skin cancer, lung cancer, kidney tumors, and bladder cancer. It may also be linked to prostate cancer and reproductive system problems.

SOURCE

Arsenic occurs naturally in the earth's crust; the western, midwestern and northeastern states are particularly rich in natural arsenic. Certain pesticides and industrial chemicals containing arsenic can also find their way into our drinking water through runoff, spills, or improper disposal. In Canada, high levels of arsenic have been found in groundwater in parts of Saskatchewan, Nova Scotia and Newfoundland.

Acceptable Levels

The Canadian Drinking Water Quality Guidelines set an interim limit of 0.025 mg/L until an appropriate treatment technique is developed to further reduce arsenic levels. Quebec, Ontario and Newfoundland have adopted this standard, while Nunavut and the Northwest Territories have regulations of 0.05 mg/L. All other jurisdictions have unenforceable guidelines or objectives of 0.025 mg/L, except Prince Edward Island and the Yukon, which lack guidelines on arsenic.

In the US appropriate arsenic levels have been the subject of considerable debate, in part because of the high cost of removing low levels of arsenic. The EPA recently settled on a limit of 0.05 mg/L, while the World Health Organization calls for 0.01 mg/L.

Treatment

Arsenic levels can be reduced to roughly 0.025 mg/L with greensand filters. Reducing levels further than this involves specialized and costly treatment such as activated alumina, ion exchange or reverse osmosis.

At home, you can reduce the level of arsenic in your tap water by using a reverse osmosis system or a distillation system that is certified to meet ANSI/NSF standards for arsenic reduction. Bottled water is also an option, but read the labels carefully to choose a brand low in arsenic.

Radon

Radon is best known as a gas that can seep into buildings from the surrounding soil and rocks, but tap water is also vulnerable to this radioactive gas.

Health Effects

Radon can cause lung cancer and cancer of the gastrointestinal tract. Smokers are particularly vulnerable to the effects of radon — radon and cigarette smoke appear to work synergistically to increase the risk of lung cancer. The National Research Council estimates that 180 Americans die each year from lung and stomach cancers caused by radon in drinking water.

SOURCE

Radon is a radioactive gas that is naturally present in certain rocks and can leach into neighboring groundwater. New England, the Appalachians, and the Rocky Mountain regions tend to have higher levels of radon in their groundwater. It's not just drinking radon-contaminated water that is dangerous — you're actually exposed to more radon through showering and bathing, which releases radon gas into the air where you can breathe it in.

ACCEPTABLE LEVELS

Although the 1996 Safe Drinking Water Act Amendments require the EPA to establish regulations for radon, these have not yet been finalized. Likewise, there are currently no Canadian guidelines for radon.

TREATMENT

If you live in an area with high levels of radon in the soil and you rely on groundwater for your drinking water, consider a filter system that meets ANSI/NSF standards for radon removal. Since so much radon is absorbed during bathing and showering, a point-of-entry filter system that treats all the water coming into the house is preferable to a point-of-use system that treats only the water at a particular tap.

Perchlorate

Perchlorate is one of a host of industrial chemicals that can contaminate drinking water. It is a key ingredient in rocket fuel.

HEALTH EFFECTS

Perchlorate affects the thyroid gland, interfering with its normal function and increasing the risk of thyroid tumors.

SOURCE

Contamination is usually caused by rocket fuel spills or leaks at military facilities so it's only an issue in certain areas. It has been found in the drinking water of Phoenix, Las Vegas, and other cities that depend on the Colorado River for their water.

ACCEPTABLE LEVELS

The EPA is currently assessing the toxicity of perchlorate. No regulations currently exist either in Canada or the US.

CHAPTER THREE

How is it Treated?

JUST AS WATER CAN BE CONTAMINATED both by natural processes and through human action, it can also be cleansed through natural means or human intervention. Some purification occurs naturally, as water percolates through soil or travels down a stream. However, these processes do not usually create drinking-quality water. For this to be achieved on a larger scale, water treatment plants are required.

THE NATURAL CLEANSING PROCESSES

As groundwater seeps through the earth to underground layers of water-bearing rock called aquifers, contaminants are naturally filtered out. How much gets removed depends on how thick a layer of earth the water passes through and what that layer is made of. Clay is the most effective at filtering water, but silt and sand are also good if they are sufficiently fine-grained and form a thick enough layer. There are also living organisms in the soil that break down organic contaminants, and any pathogens that do make it into groundwater supplies don't tend to survive — conditions in aquifers are not very hospitable to micro-organisms.

Purification also takes place in surface water, but for different reasons. There are many micro-organisms naturally present in the surface water that feed on organic contaminants and break them down. There are no processes that occur in surface water to remove suspended particles, but sunlight helps to kill off harmful bacteria and viruses.

Figure 3.1 Surface Water Treatment

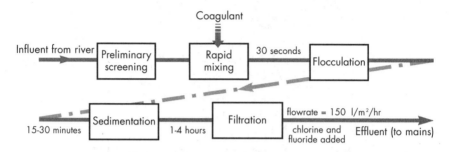

In this typical surface water treatment plant, water goes through several stages before it is ready for drinking. The complete process takes several hours.

WATER TREATMENT PLANTS

The purpose of a municipal water treatment plant is to provide water that tastes reasonably good and is safe to drink. And because many treatment plants supply water to local industries as well as to households, they also need to produce water that's clean enough for most industrial uses. To accomplish all this, treatment plants use a variety of processes to cleanse and purify water. Which processes are used depends on the quality of the incoming raw water, so the treatment plant must be flexible enough to deal with daily and seasonal changes in the raw water quality. Generally, surface water requires more treatment than groundwater because it tends to be more contaminated (see Figure 3.1, above). The most common processes are described below.

Preliminary Screening and Straining

Screening and straining are not necessary for groundwater, but they are important for removing debris from surface water. Screening removes the largest bits, such as leaves, twigs, rags and all the other miscellaneous stuff that can end up in a river or lake. The screens can be vertical bars, rotating drums, or continuous strips of perforated material located at the water intake. They must be periodically jet washed or raked to remove the debris and prevent the system from clogging up. Some treatment plants also use microstrainers made of very fine stainless steel wire mesh, designed to remove suspended solids including most plankton and algae.

Coagulation and Flocculation

Screening and straining can't remove the small particles suspended in water; that's where coagulation and flocculation come in. These processes are used to remove color, turbidity and algae. Water suppliers add a coagulating chemical such as alum (hydrated aluminum sulfate) to the water; when alum reacts with water, it forms aluminum hydroxide, which is positively charged. Most of the suspended particles are negatively charged, so they are attracted to the aluminum hydroxide and form clumps or "flocs," trapping other impurities in the process. The flocs are easily separated from the water by sedimentation and filtration. Other common coagulants include sodium aluminate, ferrous sulfate and lime, and ferric chloride.

Sedimentation or Flotation

Sedimentation (sometimes referred to as clarification) is the process of using gravity to settle flocs out of the water after coagulation and flocculation. It's a simple process — you can try it yourself by scooping up some muddy creek water in a jar and letting it sit for a few hours. You'll notice the water becomes much clearer as a layer of sediment appears at the bottom of the jar. The trick to making sedimentation happen is to make sure the water is moving slowly or is completely still — if you keep stirring the water, the particles won't settle out. For this reason, treatment plant operators use large settling tanks to reduce the speed of the water. Sedimentation can be used at the beginning of the treatment process as well, to remove suspended solids in very cloudy water.

Alternatively, gas bubbled through the tank can be used to attract flocs and transport them to the surface where they can be skimmed off. This process is called flotation.

Coagulation, flocculation and sedimentation/flotation are relatively expensive processes that require large amounts of initial capital for the tanks, and ongoing expenditures for chemicals and maintenance.

Filtration

The smallest suspended particles are removed by filtration — the process of passing water through beds of porous material. There are a large number of methods available; the two most commonly used are slow sand filtration and rapid gravity filtration.

Slow sand filtration uses layers of sand to filter the water. As the water moves through progressively finer sand, more contaminants are filtered out. Slow sand filters also have a biologically active "schmutzdecke" (a German term meaning "dirt layer"). This is a thin layer of both inorganic and organic particles derived from the raw water. This layer naturally develops on top of the sand as the water filters through (see Figure 3.2 below), and helps to break down organic particles in the water. Directly below this, a thicker layer develops, containing micro-organisms that feed on organic contaminants and potentially pathogenic organisms. Slow sand filters are simple to operate and maintain. Every few months the system must be drained and the top few centimeters of sand removed when the schmutzdecke grows too thick. A slow sand filtration system can process approximately 3.7 gallons of water per square foot per hour (150 liters per cubic meter per hour), and, unlike other large-scale filtration systems, these filters can remove non-biodegradable detergents.

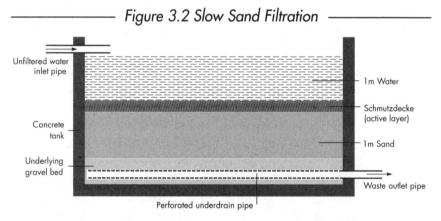

Figure 3.2 Slow Sand Filtration

In a slow sand filter, the water is cleansed by the micro-organisms in the schmutzdecke and by the sand itself. Note that the diagram is not to scale.

Rapid gravity filtration is similar to slow sand filtration, but it is much faster and can handle a flow rate of 98 gallons per square foot per hour (4,000 liters per square meter per hour). It is also cheaper to install and operate, and occupies less area. However, the sand particles in a rapid gravity filter aren't as fine as the ones in a slow sand filter, and no schmutzdecke forms, so it's not as efficient at removing contaminants. The system must also be backwashed after becoming clogged, so rapid gravity filtration requires more energy than slow sand filtration. Both methods remove quite a few bacteria along with the fine particles, and a layer of activated carbon is sometimes added to sand filters to remove organic contaminants, color, and taste- or odor-causing compounds.

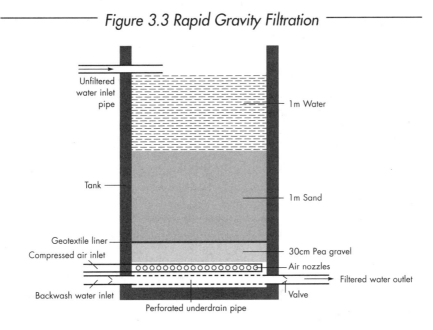

Figure 3.3 Rapid Gravity Filtration

In a rapid gravity filter, the water is cleansed as it passes through the layer of sand. The sand must be aerated and backwashed frequently. Note that the diagram is not to scale.

Disinfection

Disinfection is the most important step — and probably the most controversial one — in the treatment process. The object is to kill or inactivate all the pathogens that might be in the water, including bacteria, amebic cysts, algae, spores and viruses.

The ideal disinfectant should be effective against all pathogens present in whatever quantity are there. It should leave a "residual" — that is, some of it should still be present to continue to disinfect the water after it leaves the plant — protecting the water from possible recontamination as it travels to your tap. It should be cheap, reliable and easy to produce, and it should be fast and effective in a variety of water conditions. Finally, it should create no harmful byproducts.

Unfortunately, no single disinfectant meets all these requirements, and choosing the most appropriate one means making trade-offs. The most commonly used disinfectants are chlorine, chloramines, chlorine dioxide, ozone and ultraviolet irradiation. As government regulations become stricter, new technologies like membrane filtration are also becoming more widely used. The strengths and weaknesses of each method are described below.

If a disinfection system fails, for whatever reason, the water supplier should issue a "boil water order" advising consumers to boil their water to kill any possible pathogens. This is an excellent short-term measure to prevent waterborne disease. It's not a good long-term measure for dealing with pathogens, however, since boiling water will actually concentrate some of the other contaminants that might be present.

CHLORINE

In 1881, Robert Koch first demonstrated that chlorine could kill bacteria. This knowledge was put to use in 1905, when continuous chlorination of a public water supply was used to combat typhoid epidemics in London, England. Since then, chlorine has been widely adopted around the world as a water supply disinfectant. According to the American Water Works Association, more than 79,000 tons of chlorine are used each year to treat water in the US and Canada.

Chlorine has a lot of practical strengths. It can be administered as a liquid or as a gas, it is effective against viruses and bacteria, and it leaves traces that continue to disinfect after the water has left the treatment plant. It is reliable and relatively cheap.

It does have a few drawbacks, however. First of all, many consumers find too much chlorine gives their drinking water an unpleasant, chemical taste. Secondly, chlorination creates by-products like trihalomethanes (THMs) and haloacetic acids (HAAs), many of which cause cancer in laboratory animals. In addition, several epidemiological studies have found connections between the presence of chlorine by-products in drinking water and increased rates of certain cancers in the humans drinking it. Finally, chlorine is not effective at removing all pathogens: *Cryptosporidium* and *Giardia* oocysts tend to be resistant to chlorine disinfection. Some researchers are now suggesting that chlorine can be effective against *Giardia* if the contact time is long enough. Unfortunately, the longer the contact time, the greater the opportunity for THMs to form.

Disinfection byproducts can be minimized. Treatment plants can reduce the amount of organic material in the raw water prior to disinfecting it, or by using just enough chlorine necessary for disinfection. They can avoid the problem entirely by using a non-chlorine form of disinfection. Finally, they can remove THMs once they've been formed. Minimizing disinfection byproducts is important, but the bottom line is that the risks of not disinfecting water far outweigh the risks created by disinfection byproducts.

CHLORINE DIOXIDE

Chlorine dioxide is often used as an alternative to chlorine. This powerful oxidant controls taste and odor problems. It is effective at killing pathogens including *Cryptosporidium* and *Giardia*, and it leaves a residual disinfectant. However, like chlorine, it produces harmful chlorinated byproducts. Chlorine dioxide must be generated on site by combining chlorine and sodium chlorite. It is widely used in Europe, but is not as common in the US. It is almost ten times as expensive as chlorine.

CHLORAMINES

Chloramines are formed from chlorine and ammonia. They are moderately effective against bacteria, but not so good at killing viruses. Their advantage is that they don't break down quickly so they produce a longer lasting "residual" than chlorine — that is, they continue to protect treated water from pathogens after the water has left the treatment plant. This is a double-edged sword: the fact that they are long-lived means they protect consumers from nasty pathogens, but it also means that they're still present when tap water eventually makes its way back into lakes, rivers and oceans, and that concerns many environmentalists. Mono- and dichloroacetic can be produced as byproducts of chloramine use, and if chlorine is added prior to ammonia to form chloramines *in situ*, all the byproducts associated with the use of chlorine can be formed. Since all these byproducts of chloramines contain chlorine, they are potentially harmful. Chloramines are used widely in the US and cost roughly twice as much as chlorine.

ULTRAVIOLET IRRADIATION

Ultraviolet (UV) radiation is generated by special light bulbs which are immersed in the water. The UV rays work by damaging the genetic material in bacteria and viruses, preventing them from reproducing. It is also effective at inactivating *Cryptosporidium* and *Giardia* oocysts. The biggest advantage of UV irradiation is that it does not generate harmful by-products. Also, it requires minimal contact time so it doesn't demand the huge retention tanks that chlorine does, and it's not affected by pH. However, it leaves no residual in the distribution system, so there is no sustained disinfection. Treatment plants that use UV often add a small dose of chlorine for residual protection. The effectiveness of UV irradiation is reduced by suspended solids in the water, so it is important that the incoming water is cleaned well before it is disinfected. It is widely accepted as a disinfectant for private supplies and is now increasingly used for public supplies, particularly in Europe. In the US, there are more than 1,000 systems using UV irradiation, and several Canadian municipalities from Victoria to St. John have embraced this technology. When Seattle's new system is in place in 2004, it will be the world's largest municipal UV system, handling up to 180 million gallons of water each day. UV irradiation costs roughly twice as much as chlorine disinfection.

OZONATION

Ozone is generated on site by passing dry air or oxygen (O_2) through an electric charge, converting it to ozone (O_3). The ozone gas is then bubbled through the water. It is an excellent disinfectant for bacteria and viruses, eliminates tastes and odors, and breaks down pesticides. Ozone is one of the few disinfectants capable of inactivating *Cryptosporidium*.

However, it does create a number of by-products, including formaldehyde, acetaldehyde, and sometimes bromates (if the raw water contains bromide ion). These have been shown to cause kidney tumors in rats and may cause cancer in humans. Ozone does not produce a residual disinfectant, so another disinfectant must be added to protect the treated water from recontamination. The equipment has proved to be less reliable than other methods of disinfection. Furthermore, it can lead to higher rates of corrosion in the distribution system.

Ozone is widely used in Europe, and there are over 3,000 water treatment plants worldwide using ozone as either disinfectant or strong oxidant. In the US there are more than 400 water treatment plants that rely on ozone, including plants in San Francisco, Oklahoma City and Milwaukee. It is not as common in Canada, but ozone is used in a number of plants across the country, particularly in Quebec. Ozonation costs roughly four times as much as chorine disinfection, due mostly to the large amounts of electricity used to generate the ozone.

MEMBRANE FILTRATION

A final option for removing pathogens is the use of membrane filters. These are filters that have extremely small pores — so small that viruses, bacteria and oocysts, as well as suspended solids and many chemical contaminants, cannot fit through them. Membranes are classified by their pore size, ranging from reverse osmosis filters with pores of 0.0001 μm, to the relatively large pores of microfilters which are 0.1 μm in diameter. Currently, membrane filtration is roughly twenty times as expensive as chlorine disinfection, although the costs are steadily decreasing.

Additional Treatment Processes

Screening, coagulation, filtration and disinfection are the most commonly used water treatment techniques, but there are also a number of specialized and advanced water treatment processes that may be used, depending on the quality of the raw incoming water. These include removing specific synthetic and natural organic matter, softening hard water, and removing iron, manganese and objectionable tastes and odors. Some treatment plants add corrosion inhibitors to prevent the water from leaching lead and other metals out of pipes and plumbing fixtures. And some treatment plants add flouride to promote the dental health of consumers.

Fluoridation

Fluoride protects teeth against cavities, up to a certain limit. The higher the level of fluoride you ingest, the fewer cavities you're likely to develop. It works by increasing the resistance of your tooth enamel to the acids produced by bacterial plaques. If fluoride levels are too high, however, dental fluorosis can occur. This is a condition where teeth become mottled and discolored. In extreme cases the enamel becomes pitted. The optimal concentration of fluoride to maximize protection against cavities and minimize the risk of dental fluorosis is 1.0 mg/L.

Most waters naturally contain some fluoride, but the concentration varies a lot depending on the water source. Some communities therefore supplement the natural levels by adding fluoride, and some defluoridate their water because natural levels are too high.

Early studies showed that fluoridating water to 1.0 mg/L had great benefits, reducing the prevalence of cavities by 40 to 50 percent in baby teeth and by 50 to 60 percent in adult teeth. More recent studies have shown much less difference in the rates of tooth decay between communities with fluoridated water and communities without, probably because of the widespread use of fluoridated toothpaste.

The first community water fluoridation scheme was set up in the United States in 1945. In 1969, the World Health Organization recommended that countries introduce community water fluoridation because it was (and still is) the least expensive and most effective way of providing fluoride to large numbers of people. Today, more than 60 percent of the tap water in America is either naturally fluoridated or has fluoride added to it, and the Centers for Disease Control would like to see that number increase to 75 percent by 2010. Many respected health associations have endorsed fluoridation, including the American Dental Association, the American Medical Association and the US Public Health Service.

Fluoridation may be good for your teeth, but there are a number of critics who worry that fluoride increases the risk of osteoporosis and certain types of cancer. The current scientific consensus is that there is no link between fluoride and cancer. The evidence regarding osteoporosis is a little less clear — most studies show fluoride has no effect on bone health, a few suggest fluoride increases the risk of osteoporosis, and a few actually suggest it has a mild protective effect. Unfortunately there is a shortage of well-designed research studies on this issue.

There is no doubt that high levels of fluoride can affect the nervous system and can cause skeletal fluorosis, a condition similar to osteoporosis; however, these doses are far, far higher than the levels added to drinking water. There's also no doubt that tooth discoloration from too much fluoride is more common in North American in the past 50 years, although most of these cases are considered "mild" or "very mild." Fundamentally, for a number of people the issue of community fluoridation comes down to an issue of choice. Once a treatment plant adds fluoride to a community water supply, anyone who objects to it can't boil or filter it out. Community fluoridation is not the only way to administer fluoride — if you want fluoridated water, you can add fluoride tablets to your own drinking water while your neighbors can choose to drink fluoride-free water.

SPECIAL CHALLENGES FOR SMALL SYSTEMS

While many water treatment systems in large cities provide specialized treatment, have highly trained staff, and monitor water quality on a daily basis, that's not always the case in smaller towns. Increasingly, governments in North America are waking up to the fact that treatment systems serving small communities face some special challenges.

First, small treatment systems don't have access to the same resources — human or financial — that larger systems do. Average incomes in smaller communities tend to be lower than in larger centers, and they are home to fewer businesses and industries. These factors mean a smaller tax base, making it more difficult for small treatment systems to afford the equipment or the qualified operators necessary for high quality treatment. An EPA survey found that many small community water systems were in poor financial shape, and more than 30 percent had no money in reserve to repair or upgrade their facilities — they had enough money to take care of daily expenses, but nothing more.

Second, small systems are often more vulnerable to source contamination, both because they have fewer water sources to choose from and because they are frequently located in areas affected by logging, agriculture, mining, grazing, or other land uses that can degrade the quality of source water.

In British Columbia, there are potentially serious problems where small community water systems have no identified supplier — no one treating the water, no one maintaining the system and no one monitoring water quality. The situation can arise when a developer builds a water supply system for a new subdivision, and then walks away from it once the houses have been sold. In other cases, homeowners allow neighbors to tap into their private wells. Some of these so-called "orphan" or "good neighbor" systems may have up to 80 connections.

THE MULTIBARRIER APPROACH

The water treatment plant is the core of any good drinking water system, but it's not the only component. Case after case has proven that treatment plants can fail — because of human error, mechanical breakdowns, or extraordinary conditions. When that happens, additional barriers can mean the difference between a safe water supply and a serious outbreak of waterborne disease.

One additional barrier is good source protection, based on the old adage that prevention is the best medicine. The cleaner the raw water, the lower the risk if the treatment plant fails for any reason. (See Chapter Five for more about source protection.) The other barriers focus on the water once it has left the treatment plant. A reliable, secure distribution system (see Chapter Four) ensures that treated water isn't recontaminated on its journey from the treatment plant to your tap. Regular testing (see Chapter Six) provides assurance that the treatment plant and the distribution system are working effectively, and alerts the authorities when they're not. Finally, a well rehearsed plan to deal with any contamination that does occur is vital. That includes a system for reaching every member of the community to let them know what the problem is and what steps they need to take to protect their health.

CHAPTER FOUR

The Distribution System

MODERN WATER TREATMENT and distribution is a massive engineering feat. Thousands of kilometers of pipes carry water from distant sources to centralized plants where it is treated, and then to individual homes, where clean water flows out of taps on demand. There are approximately 1 million miles (1.6 million kilometers) of pipes and aqueducts in the United States and Canada — enough to circle the Earth forty times — and a single large city may depend on a network of 600,000 miles (1,000,000 km) of pipes to distribute drinking water.

First, raw water has to get from the source (either groundwater or surface water or both) to the treatment plant. Usually the plants are located close to the source, but this is not always the case. For example, the city of Colorado Springs depends on water sources nearly 200 miles (325 km) away to supply most of its needs. Raw water is collected from mountain springs and stored in reservoirs along the Continental Divide, then transferred through a gravity-fed system of pipes and reservoirs to Colorado Springs. Winnipeg, Manitoba pipes its water from Shoal Lake, approximately 90 miles (150 km) east of the city.

Once it has been treated, the next challenge is to deliver that water to consumers. This is done through the distribution system — an underground maze of pipes and tanks that extends for scores of kilometers, regulated through strategically placed pumps and valves that allow water suppliers to maintain and repair the system. Let's look at that system.

After treatment, water is usually stored in service reservoirs. Reservoirs serve two functions: they store enough water to meet peak demand, and, if they are raised up, they create water pressure in the mains. Sometimes water towers are used to create pressure in very flat areas. From the reservoir, the water travels through a network of trunk mains, which take water from service reservoirs and supply it to feeder mains. Next, the feeder mains take the water from trunk mains and deliver it to focal points within the distribution network. Finally a network of small mains distributes water from the feeder mains to consumers' supply pipes. This is where the distribution system ends and, with it, the water supplier's responsibilities. Until it reaches your supply pipe, however, the water supplier must ensure the water stays clean and safe to drink.

That's not always easy, for a number of reasons. First, the longer the water remains in the system (particularly if it stands still), the more its quality declines. A good system avoids dead ends as much as possible; in fact, there are often loops in the distribution system to prevent water from stagnating.

Second, water in the distribution system can become contaminated. This is a major cause of waterborne disease: anywhere from 30 to 78 percent of waterborne diseases can be traced to a contamination problem in the distribution system. For example, in 1995-1996, 70 percent of the outbreaks associated with community water systems in the US were caused by problems in the distribution system or in customers' plumbing. These problems can include corrosion, cross-connections, backflow, breaks and leaks, and buildup of biofilm.

CORROSION

Most mains are made of iron or steel and are therefore vulnerable to corrosion. Lead components are still present in approximately 20 percent of public water distribution systems in the US, which creates the possibility of lead leaching into the treated water as the pipes corrode. As discussed in Chapter Three, the water supplier can minimize this risk by adding anti-corrosion agents to the water, and by using pipes with anti-corrosion coatings.

In addition to increasing the level of metals in the treated water, corrosion can reduce the life of water distribution pipes by weakening the walls and increasing the likelihood of leaks, breaks and contamination.

CROSS-CONNECTIONS AND BACKFLOW

A cross-connection is any place in the system where the treated water can come into contact with wastewater, sewage or untreated water.

Cross-connections are the leading cause of contamination in the distribution system.

Where cross-connections exist, there is a risk of backflow — a situation where the flow of the treated water reverses, and untreated water is sucked into the distribution system. This can occur either because the pressure in the water mains drops, or because the pressure in the untreated source becomes greater than the pressure in the water mains. Any situation where large amounts of water are being withdrawn can lead to backflow by causing a pressure drop in the mains: when a water main breaks, for example, or when firefighters are using high volumes of water from fire hydrants.

Usually the contamination that results is quite localized: a building, a few houses or an apartment complex, for example, but occasionally it can affect a wider area.

Backflow can be prevented though the use of one-way valves, but because most cross-connections occur within residential, commercial, institutional and industrial properties, it is difficult for water suppliers to identify and check them. Many water suppliers work with building inspectors and public health agencies to detect and deter cross-connections, particularly on sites where backflow could be

Examples of Backflow

The EPA has compiled a list of backflow incidents in the United States from 1970-1999. Here are just a few examples:

- *In 1981, pesticides were backsiphoned into an apartment building in Pennsylvania, affecting 75 apartment units. The culprit was a garden hose connected to a termite exterminator's tank truck filled with chlordane and heptachlor.*

- *In 1997, 40,000 neighborhood taps in Charlotte-Mecklenburg, North Carolina were contaminated by firefighting chemicals. The incident occurred when a fire truck pump created backpressure on a fire hydrant before the valve was closed. As a result, 60 gallons (230 liters) of foam were forced into the distribution system.*

- *In 1990, more than a thousand guests at a country club in Tennessee came down with intestinal disorders after consuming drinking water from a well contaminated with sewage from a malfunctioning sewage pumping station.*

- *And more recently, in Tennessee, cross-connections between treated water pipes and pipes carrying untreated creek water were discovered at an atomic bomb fuel plant. Tests showed the creek water was contaminated with strontium-90 and arsenic. It is not known how long the cross-connections were in place — possibly decades.*

particularly dangerous: hospitals, funeral homes, and food packing plants, for example. Pressure-monitoring equipment can also alert a water supplier when a pressure drop occurs.

BREAKS AND LEAKS

Breaks in water distribution pipes are not uncommon. According to one study, in water systems serving more than 500,000 people, there are an average of 488 breaks each year in the mains. Even well run water distribution systems experience 25 to 30 breaks per 100 miles of piping per year (16 to 19 breaks per 100 km). Breaks are a particular concern in older distribution systems, as aging pipes become more susceptible to frost, traffic vibrations, the erosion of supporting ground materials, and earth tremors.

Aging systems are likewise more susceptible to leaks. Many cities in North America are faced with distribution systems where some components can be more than a century old. As a result, they lose more than 20 percent of treated water in the distribution system. In addition to wasting water, leaks can undermine the supporting ground material, further increasing the risk of breaks and bursts. And, of course, leaks can allow contaminants into the distribution system.

BIOFILM

Biofilm is a film of bacteria that can grow on the walls of distribution pipes. These bacteria feed on any organic material that isn't removed from water during the treatment process. Some types of bacteria even thrive on iron and sulfur, which can be found in groundwater — and the result is very unpleasant-smelling water!

Biofilm can contaminate treated water, which is why it's important for water suppliers to make sure there is residual disinfectant in the water as it travels through the distribution system. Unfortunately, residual disinfectant alone may not be enough to keep biofilm under control, so water suppliers need to regularly monitor pipes and flush them if biofilm becomes a problem. Biofilm can also accelerate corrosion — another reason for water suppliers to keep it under control.

CHAPTER FIVE

Protecting the Source

COMMUNITIES HAVE TRADITIONALLY put a lot of money into water treatment, monitoring and distribution systems. Today, many communities are also starting to invest in another critical component of good drinking water systems: source protection.

All kinds of things can contaminate source water. Cattle grazing near a surface water source can lead to fecal contamination; logging in a catchment area increases the amount of sediment that washes into lakes or rivers; and the pesticides used in industrial forestry can run off into streams or leach into groundwater. Likewise, the pesticides and fertilizers that farmers spray on their crops can threaten nearby streams and aquifers. Other common sources of contamination include industrial wastes, mining, septic tanks, and sewage treatment plants.

PREVENTION PAYS

It makes good sense to keep water sources safe from contamination — good sense both from an economic perspective and from a health perspective. Let's look at the money angle first.

There are many contaminants that standard treatment techniques cannot remove. For example, it takes expensive ion exchange processes to remove excess nitrates, a common contaminant in agricultural areas. It's cheaper and easier in many cases to prevent nitrates from getting into water sources in the first place.

The same principle is true for many industrial chemicals, so having strict regulations on what industries can and cannot discharge in their wastewater helps ensure good drinking water. By the time water reaches the treatment plants, it can contain a cocktail of unknown substances from thousands of sources. It is far easier to remove contaminants at source, where it may even be possible to recycle them.

High quality source water means less treatment is required: water with low levels of turbidity, bacteria, viruses and other contaminants can be treated successfully with just filtration and disinfection. The more contaminated the source water, the more treatment is required and the higher the cost of producing safe drinking water.

Finally, consider the cost of waterborne disease caused by poor source protection. Cranbrook, BC suffered an outbreak of cryptosporidiosis in 1999 that was traced to cattle grazing within the watershed. The Cranbrook Chamber of Commerce estimates the city lost Can$5 million (US$3.6 million) in business and tourism revenue as a direct result of the outbreak.

From a health perspective, source protection is an important component of a multibarrier approach to safe drinking water. Treatment plants aren't always effective 100 percent of the time, so we need to have these additional safeguards in place. A good multibarrier approach begins with source protection, because the cleaner the source water, the lower the risk of disease outbreak should a treatment plant fail for any reason.

COMMUNITY SOURCE PROTECTION

The first step in developing an effective source protection program is to identify the boundaries of the area that can influence the quality of the source water. This can be fairly straightforward if you're talking about surface water, but it's a lot more challenging when it comes to groundwater. Next, communities should identify what potential contaminants lie within that area: where they are, what they are, and how much of a threat they pose. Are there animal feedlots in the watershed? Leaking underground storage tanks? Abandoned wells? Heavy industry? Once this initial assessment is done, communities can decide on the most appropriate way to tackle the biggest threats.

There are several ways municipalities can protect their source water. In some cases, they can buy the land surrounding surface water sources or wells to ensure it is protected from any dangerous activities.

Contra Costa County in California has chosen this route, by purchasing 99 percent of the land within the Los Vaqueros Reservoir watershed, giving the water district maximum control over land use in the area. The water district continues to allow sheep and cattle grazing in the watershed, provided it complies with best management practices, but it has established a fenced buffer area of 100—300 feet (30—90 meters) surrounding the reservoir, as well as all the major watercourses that feed it. Angling is permitted along the shoreline, but the only boating allowed is restricted to electric-powered boats owned and operated by the water district, thus avoiding the risk of contamination from fuel spills and leaks.

In other cases, municipalities can restrict the type of land use that is permitted near water sources, upgrade their sewage treatment plants, or pass stricter bylaws on waste and wastewater disposal.

In British Columbia, the city of Vancouver has banned logging in the three forested watersheds that provide the city's drinking water. Between 1967 and 1999, almost 5,000 hectares (2,000 acres) of forests were logged and 300 kilometers (185 miles) of roads were constructed. The result was increasingly turbid water caused by soil washing away once the tree cover had been removed, and the city was faced with higher and higher costs for water filtration and chlorination. In 1999, the logging company's licence for the area was cancelled, and the city has established a no-logging policy on watershed lands. Victoria goes even further, prohibiting people from hiking on watershed lands.

In New Brunswick, all communities that rely on surface water sources are protected by a Watercourse Setback Designation Order, which severely restricts land use within 75 meters (0.5 miles) of the lake, river or stream in question. The Order also limits agriculture, forestry, road construction, mining and commercial and industrial development within the watershed as a whole.

Agricultural runoff is a big issue in many communities. Burlington, Vermont has worked with local farmers to resolve the issue of spring runoff from fields contaminating Lake Champlain, the city's source of drinking water. Farmers agreed to a moratorium on spreading manure on frozen fields, which is much more likely to wash away in the spring melt.

One of the most ambitious source protection programs in North America is taking place in New York. About nine million people rely on New York City drinking water, which comes from surface water sources upstate: the Croton and Catskill-Delaware watersheds. Traditionally, the city hasn't filtered this water, relying on chlorination as the only form of treatment. However, under the EPA's 1989 Surface Water Treatment Rule, it was faced with the choice of either installing filtration systems at the cost of roughly seven billion dollars, or proving to the EPA that its drinking water sources met stringent protection standards. It chose the second option, tackling the issues of farm runoff, sewage treatment plan discharges, and pollution from local residences and businesses.

The biggest challenge for New York City was balancing the need to protect its watersheds with the needs of the towns, villages and farms within the area. Its approach is multi-pronged. To deal with the issue of farm runoff, it created the voluntary Watershed Agricultural Program which helps farmers implement best management practices to prevent pesticides, fertilizers and soil from washing into local streams. To date, more than 90 percent of local farmers have signed up to participate. On other fronts, the city poured money into fixing leaking septic tanks, constructing stormwater detention ponds, and upgrading substandard sewage treatment plants within the watersheds. So far, the results have been excellent and the city is having no problem meeting the EPA's standards.

STATE AND PROVINCIAL LEGISLATION

States and provinces can also pass legislation to protect drinking water sources. Under the 1996 Amendments to the US Safe Drinking Water Act, each state must develop a program to identify potential contamination threats. Currently, over one-third of community water systems in the United Sates have some kind of protection program in place.

Source protection is particularly important when it comes to groundwater, because once an aquifer has been contaminated, it is extremely difficult, costly and time-consuming to clean it up. Tallahassee, Florida has been forced into an expensive cleanup of seven downtown wells because they became contaminated with solvents from local dry cleaners and businesses that degrease machine parts. To prevent future problems, Tallahassee launched an aquifer protection program that involves frequent inspections of local businesses, registration of businesses that use or manufacture potential contaminants, and stricter regulations on waste disposal.

San Antonio, Texas has focused on identifying and closing abandoned wells, because they are a direct conduit to groundwater supplies. Surface runoff can get into abandoned wells, carrying with it pathogens, fertilizers, pesticides and other contaminants. In the case of San Antonio, contamination via abandoned wells was having a significant impact on the aquifer that supplies the city's drinking water.

The EPA has developed two protection programs that target groundwater. Under the "Sole Source Aquifer Program," communities, organizations and even individuals can petition the EPA to designate an aquifer that is the principal or sole source of drinking water for the local population. Once an aquifer has been designated a sole source, the EPA has the authority to review any projects financed with federal money that have the potential to contaminate the aquifer.

The EPA has also developed a voluntary Wellhead Protection Program. Under this scheme, 49 states and territories have developed and implemented programs to protect land surrounding wells from anything that could contaminate the underlying groundwater. Eleven states have also developed comprehensive groundwater protection programs that include wellhead protection, hazardous waste management, and pesticide measures.

In Canada, several provinces have created legislation to protect wellfields and watersheds. Newfoundland has been particularly proactive in protecting its water sources, and 70 percent of the population now receives their drinking water from designated Protected Water Supply Areas. New Brunswick is in the process of applying its new "Wellfield Protected Area Designation Order" to communities that rely on groundwater. This Order severely restricts potential pollutants in a zone around the wellhead — for example, livestock are prohibited and manure cannot be stored or spread in this area.

An important component of many source protection programs is education: telling citizens, businesses and farmers how their activities can damage source waters and how they can adopt best management practices. For example, many cities across North America have run very successful storm drain marking programs, making the public aware that putting oil, paints and other hazardous wastes down storm drains harms the environment and can contaminate drinking water sources.

Chapter Six

Testing Drinking Water

REGULAR TESTING is the final component of an effective, multi-barrier approach to safe drinking water. It's the proof that source protection, treatment and distribution are doing their job — providing high-quality drinking water. And when something goes wrong, regular monitoring will ensure that water system operators find out quickly, so that they can respond quickly.

Everyone agrees that testing is important, but there's lots of disagreement when it comes to the nitty-gritty: which contaminants need to be monitored? How often do they need to be monitored? Who should do the testing? Who needs to be informed of the results?

WHICH CONTAMINANTS?

There's no question that water suppliers need to monitor pathogens, since bacteria, viruses and parasites in the water system can cause acute disease outbreaks. But what about physical and chemical contaminants? Which ones are worth spending the money to monitor?

The United States takes a comprehensive approach: all water suppliers must test their treated water for a list of 80 microbes and physico-chemical contaminants. Some states have gone even further than the EPA regulations. New York, for example, demands that water suppliers test for a number of organic chemicals not regulated by the EPA.

Other jurisdictions, such as British Columbia, have chosen a more selective approach, putting their resources into testing only for contaminants that have a reasonable likelihood of showing up in the system.

HOW OFTEN?

In general, surface water needs to be tested more often than groundwater, because surface water is more vulnerable to contamination, and the quality can change more quickly.

The EPA has developed a monitoring schedule for each contaminant, which depends on the size of the water supplier and the source of the water (see Figure 6.1, below). If a water supplier can demonstrate its water has been free of certain contaminants over a specified period of time, it can receive permission to test less frequently.

Figure 6.1 Sample Monitoring Schedule from the EPA

Contaminant	Minimum Monitoring Frequency
Bacteria	monthly or quarterly, depending on system size and type
Protozoa and viruses	as indicators of protozoa and bacteria: continuous monitoring for turbidity and monthly monitoring for total coliforms
Nitrate	annually
Volatile organics (e.g., benzene)	groundwater systems: annually for two consecutive years surface water: annually
Synthetic organics (e.g., pesticides)	larger systems: twice in three years smaller systems: once in three years
Inorganics/metals	groundwater systems: once every three years surface water systems: annually
Lead and copper	annually
Radionuclides	once every four years

(Source: Water on Tap: A Consumer's Guide to the Nation's Drinking Water, EPA, 1997.)

As with the American regulations, Canadian guidelines vary depending on the size of the population: the larger the community, the more frequently the water should be tested. Alberta is the exception. It insists that all water suppliers follow the same regulations, regardless of size, to ensure that everyone in the province has access to the same quality of drinking water.

BY WHOM?

In Canada, most jurisdictions insist that water quality testing be done in certified labs, provincially selected labs, or government-run labs, but this isn't the case everywhere: Prince Edward Island and Nunavut, for example, currently don't require labs to be certified. The US insists that all testing be conducted in federally certified laboratories. Testing and reporting can be audited by state or federal agencies, but in the vast majority of cases it's done on the honor system.

Accreditation ensures that labs have trained staff and the appropriate equipment and procedures to test drinking water. Essentially, it's an assurance of competence. Canadian studies comparing accredited and non-accredited labs found non-accredited labs were far more likely to perform poorly on proficiency tests and provide inconsistent results.

WHO SHOULD KNOW THE RESULTS?

And finally, who should know the test results? In the US, the state government must be informed of the results. This isn't always the case in Canada. In some jurisdictions, the lab must inform the local medical health officer; in others, it's the Ministry of the Environment that needs to be informed. In British Columbia, both the water supplier and the regional medical health officer must receive test results.

Tragically, in the case of the *E. coli* outbreak in Walkerton, Ontario, the private lab that performed the water quality testing was not required to inform anyone of the adverse test results except the water supplier, and the water supplier kept the results secret and failed to act on them for several days, greatly magnifying the extent of the outbreak.

Both governments and citizens have a lot to gain from knowing what's in our drinking water. By compiling data from different water suppliers, governments can monitor regional and national trends and use this information to identify common problems and set appropriate regulations. The US has an excellent data collection system where states compile water quality data from local water suppliers, and the EPA in turn compiles data from all of the states. Unfortunately, this is not the case in Canada.

In many provinces, water quality data is collected in a patchwork manner and there are large gaps in the information available. In British Columbia, for example, the Provincial Health Officer's 2001 report points out major deficits in the data: statistics are neither available on the number of systems that treat the water nor on the number of water systems with certified operators. Nunavut has no formal system for collecting or analyzing water quality data.

The United States is also far better at informing citizens about their local drinking water quality. One of the major amendments to the Safe Drinking Water Act in 1996 was the "right to know" provision that requires water suppliers to inform customers within 24 hours of any violation that could seriously affect their health. In addition, water suppliers must provide their customers with annual reports on drinking water quality. These "Consumer Confidence Reports" must include the following information:

- The source of the drinking water.
- The susceptibility of the source to contamination.
- The levels of contaminants in the drinking water, compared to the EPA's maximum contaminant level.
- The potential health effects of any contaminants that exceeded the EPA's maximum contaminant level.
- A description of the actions that have been taken to restore safe drinking water, if any maximum contaminant levels were exceeded.
- Compliance with other regulations.

In larger centers, consumers should receive these reports in the mail; smaller utilities can use other methods to inform their customers. If you didn't receive a Consumer Confidence Report, call your water supplier or check its website. You can also access many reports through the EPA's drinking water website (see Resources).

Unfortunately, similar regulations in Canada are few and far between. A few provinces are enacting requirements. Newfoundland publishes an annual drinking water quality report for the province, and communities receive reports on their own drinking water. Saskatchewan is just beginning to publish drinking water test results on the Internet and is planning to compile an annual province-wide report, and the Northwest Territories has established an online database of water quality test results. Ontario is at the forefront of consumer reporting with its new requirements for quarterly reports. In addition, the government will be required to produced an annual "State of Ontario's Drinking Water Report."

Elsewhere, a few water suppliers provide their customers with annual water quality reports on a voluntary basis, such as Victoria, Vancouver and Edmonton.

CHAPTER SEVEN

How is it Managed?

REGULATING WATER

IN THE UNITED STATES, public drinking water quality is regulated under the Safe Drinking Water Act, which is administered by the federal Environmental Protection Agency. Under the Act, the EPA has established standards for 80 chemical and microbial contaminants. Water suppliers are not allowed to exceed these "maximum contaminant levels." The EPA has also established how often water suppliers must test their water for each of these contaminants.

In most cases, the EPA regulations are enforced by state governments — usually by the state health department. State officials certify water system operators, conduct on-site inspections, and make sure water suppliers meet EPA requirements. If water suppliers fail to meet EPA regulations, it's also the state officials who are responsible for taking action against the supplier and informing the EPA of the violation.

In Canada, the situation is far more complex. Because drinking water falls under provincial jurisdiction, each province has its own drinking water standards. Although Health Canada has established national guidelines for safe drinking water, it is up to each province to decide whether or not to adopt these guidelines and whether or not to make them legally binding. As a result, there are a patchwork of different regulations across the country.

To further complicate the situation, drinking water responsibilities are usually shared between several provincial ministries, rather than focused in a single, stand-alone agency responsible for protecting all aspects of drinking water.

SETTING STANDARDS

With thousands of potential contaminants out there in the environment, deciding which ones should be regulated is the first challenge for governments.

The EPA looks at how often a particular substance turns up in American drinking water supplies and how dangerous it is to human health in order to decide whether it makes sense to regulate it. Once a list of "priority contaminants" has been developed, the next challenge is to decide what limit to set for each contaminant — how much can safely be present in our drinking water.

The EPA begins by establishing a goal based strictly on health criteria: the Maximum Contaminant Level Goal or MCLG. This is the level that does not present any health risk, even to the most sensitive consumers such as infants, children, pregnant women, the elderly and people with compromised immune systems. In order to arrive at this goal, the EPA conducts a risk assessment. For substances that could cause cancer, it sets a limit that results in a risk of getting cancer between one in 10,000 and one in 1,000,000. For other substances, it sets a level that ensures no risk of harm, based on a lifetime of exposure.

In making these calculations, the EPA assumes the average adult weighs 154 pounds (70 kilograms), drinks 8.5 cups (2 liters) of water a day, and lives to the age of 70. Children are assumed to weigh an average of 22 pounds (10 kilograms) and consume 4.25 cups (1 liter) of water per day.

Unfortunately, these Maximum Contaminant Level Goals are not always realistic goals. Sometimes the technologies do not exist to reduce a contaminant to the MCLG, and sometimes the technologies exist but are prohibitively expensive. In this case, the EPA will consult with water suppliers and conduct a cost/benefit analysis to determine a legally binding Maximum Contaminant Level (MCL) as close as possible to the MCLG.

Setting standards is a time-consuming process, and the EPA has been criticized for not moving quickly enough to establish standards for a number of priority contaminants, such as radon, perchlorate, and various pesticides and industrial chemicals.

The process works quite similarly in Canada, where drinking water quality guidelines are set by Health Canada's Federal-Provincial Subcommittee on Drinking Water. The committee establishes a list of priority contaminants for review by Health Canada scientists. The scientists begin by determining what level of contaminant produces no negative health effects — the No Observed Adverse Effect Level or NOAEL. Using this information, they then calculate the maximum daily intake that consumers can safely drink — the Tolerable Daily Intake or TDI.

Finally, Health Canada scientists calculate a Maximum Acceptable Concentration (MAC) that takes into account average body weight, average drinking water consumption, exposure from other sources, and a hefty safety margin that ensures MACs are 10 to 10,000 times below the NOAEL.

The major difference between American and Canadian standards is that American standards are legally binding across the country. In contrast, the Health Canada Guidelines are simply that — guidelines — and it is up to each province to decide which, if any, of the guidelines to adopt, and whether or not to make them legally enforceable. The result is a wide variation across the country: Quebec and Alberta, for example, have very strict, legally enforceable regulations that closely parallel EPA regulations. Prince Edward Island has not formally adopted any of the Canadian Guidelines, although it is currently in the process of developing regulations that will require water suppliers to monitor several microbiological, chemical and physical contaminants.

PAYING FOR WATER

The price you pay for your tap water varies considerably across the continent. In Canada, for example, water is cheapest in Quebec, Newfoundland and British Columbia, and most expensive in the Prairies and northern Canada. The average Canadian household pays anywhere between Can$15 and $90 (US$11-66) per month. In the United States, the average household pays less than US$30 (Can$41) per month for water and public utilities, although the cost varies greatly from city to city. Customers in Seattle, for example, fork over an average of US$26.65 (Can$36.34) per month, while in Las Vegas, the bill comes to only US$12.42 (Can$16.94). If your water comes from a smaller water supplier, chances are you'll be paying more than someone living in a bigger center. Regardless of where you live, what you pay probably doesn't cover the full cost of providing safe drinking water.

Water may be free when it falls from the sky, but collecting, treating and distributing it costs money. The biggest expense is treatment, particularly if it involves coagulation, sedimentation and filtration (see *Water Treatment Plants,* Chapter Three). The poorer the quality of the raw water, the more expensive the treatment process will be. Likewise, the stricter the water quality regulations, the more it costs to meet them. The EPA estimates it costs about US$2 per 1,000 gallons (Can$0.72 per 1,000 liters), on average, to provide drinking-quality water.

Rates in most areas have increased over the past couple of decades, and consumers can expect water costs to continue to rise as many jurisdictions enforce stricter regulations, and as municipalities across the continent are forced to replace aging infrastructure.

A Canadian study estimated that underpricing water led to unmet water and wastewater infrastructure costs totaling Can$38—49 billion dollars (US$28—36 billion) in 1996 alone. Similarly, an EPA survey of community water systems in 1995 predicted that US$38 billion (Can$52 billion) would need to be invested in infrastructure over the following 20 years.

In the United States, the rates charged by private water suppliers are controlled by state public utility commissions, which look at the supplier's revenues, expenses, financial outlook and quality of service in determining whether a rate hike is acceptable or not. In Canada, the rates are set by provincial and municipal officials.

Who Pays What?

North Americans pay less than half of what consumers in the Netherlands and the UK pay, and a third of what the Danish and Germans do.

According to a survey of 14 developed countries in Europe, North America, Africa and Australia, the average cost of water is 76.4 US cents per cubic meter (see Figure 7.1). The 2001 survey, conducted by NUS Consulting Group, also found the average price of water had risen in every country; an average increase of 3.8 percent. The exception was the Netherlands, which saw a decrease in cost of 0.8 percent.

Water Metering

Although most Americans have metered water, in Canada just under half the population is charged a flat rate for water rather than being billed for the amount of water actually used. This means there's no incentive to reduce water consumption — even if the tap runs all day long, the water bill at the end of the month doesn't change.

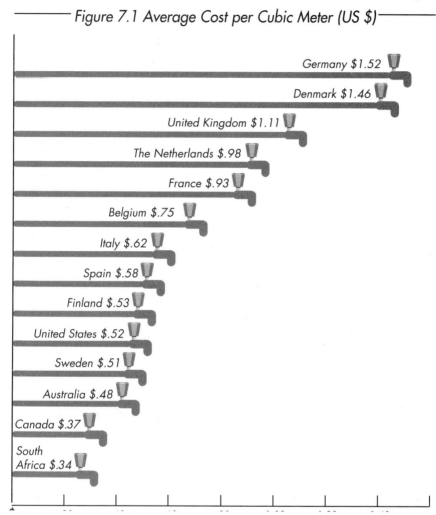

Figure 7.1 Average Cost per Cubic Meter (US $)

Not surprisingly, unmetered households use more water than households with water meters — a whopping 70 percent more, according to Environment Canada. Calgary, Alberta provides an excellent example. In 2001, 57 percent of residential accounts switched from flat rate to metering, part of an ongoing program aimed at water conservation. Calgary Waterworks recently finished a two-year study comparing water consumption between metered and unmetered customers. The results? Flat rate customers used 1.49 times more water than metered customers.

The water savings achieved by metering has led to a 33 percent drop in Calgary's peak day demand: from 1,435 liters (379 gallons) per capita in 1987, when only 21 percent of accounts were metered, to 955 liters (252 gallons) per capita in 2001. Since maximum day demand determines the size of water treatment system required, this reduction means that Calgary hasn't been forced to restrict water use during hot, dry summers, nor has the city needed to increase the capacity of its water treatment plants. The city estimates that 100 percent metering would reduce peak day demand still further to 721 liters (190 gallons) per capita, and Calgary Waterworks is currently investigating different ways to accomplish universal metering.

In 1999, 90 percent of households in Saskatchewan, the Northwest Territories, Manitoba and Nova Scotia were metered, but metering remains extremely rare in Prince Edward Island and Newfoundland. The rate is much higher across the US, where, on average, 98 percent of households are metered.

Figure 7.2 Methods of Charging for Water

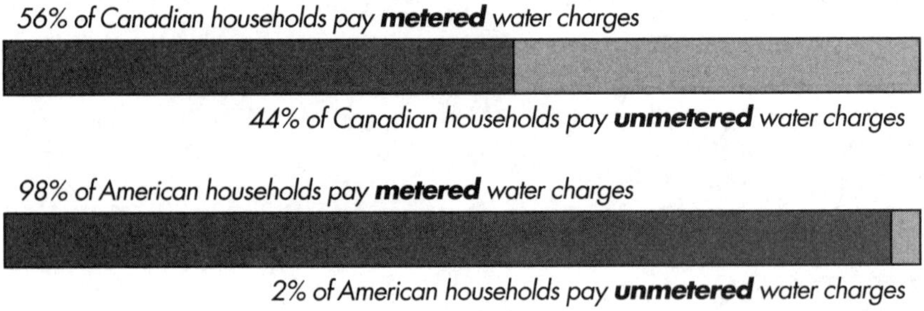

56% of Canadian households pay **metered** water charges

44% of Canadian households pay **unmetered** water charges

98% of American households pay **metered** water charges

2% of American households pay **unmetered** water charges

PRIVATIZATION

In the US, about a third of the country's water suppliers are privately owned — mostly as small, locally owned utilities. In terms of market share, private water companies accounted for 14 percent of drinking water revenues in 1995. In Canada, the numbers are much smaller — only a handful of suppliers are privately owned or operated.

However, in both countries there is increasing pressure on municipalities to privatize at least some portion of their water systems. Many water suppliers need cash infusions — which private companies can often provide — in order to upgrade aging infrastructure and invest in new technology to meet stricter water regulations. Private water companies also promise to run water systems more efficiently, and they generally succeed. However, short-term efficiencies may lead to longer term problems, if private companies choose to save money by reducing maintenance and upkeep, for example, or compromise their ability to respond to emergencies by laying off too many staff.

Throughout much of the rest of the world, privatization is the norm. Today, private companies supply water in 56 countries, and the number is rising. It's a market driven by the quest for profits; currently, the water supply business is worth about US$500 billion (Can$682 billion) a year, and industry analysts say it could easily grow into the trillions over the next few years. There are three key players in this game: the French companies Suez and Vivendi Universal, and the German conglomerate RWE AG. Together, they dominate the world market, and they are busy making inroads into North America.

In the United States, new laws have opened up the possibility for greater private sector involvement in publicly owned systems, while in Canada the shifting of responsibilities from provincial governments to municipalities has left many municipalities strapped for cash and ill-equipped to deal with these new responsibilities.

Water companies are viewing North America as a market with huge potential, and they are aggressively wooing cities with promises of significant savings, better performance and greater efficiency. They're also lobbying the US government for more favorable regulations, and they're making political contributions in the millions of dollars in a clear bid to influence government policies.

Private sector involvement can take several forms. The most straightforward is outright private ownership. In the US, about 4,000 water utilities are investor-owned, but the vast majority of these are small systems serving 3,300 people or fewer. In Canada, Edmonton, Alberta's water department is now a private company — the only one in the country.

What's more common is some kind of public-private partnership: private construction, operation and maintenance; private operation and maintenance of a publicly constructed facility; or private financing. Let's look at these in more detail.

Private Construction, Operation and Maintenance

Under this arrangement, a municipality finances the construction of a new system, but contracts a private company to do the actual building, operation and maintenance for a specified period of time. This is also known as "design, build and operate." Seattle, Phoenix, Houston and Tampa have all gone this route.

Private Operation and Maintenance

The most popular arrangement is for a municipality to contract out the operation and maintenance of an existing system to a private company, usually for a period of five years or less. This option is often referred to as "outsourcing."

There are hundreds of examples in the US, including Jersey City, Milwaukee, Indianapolis and, until recently, Atlanta. The Ontario municipalities of London, Haldimand-Norfolk and Goderich have signed this kind of contract, as has Fort Saskatchewan in Alberta.

Private Financing

The final option is for a private company to provide the financing for a new water system or upgrades to an existing system. In order to recoup their investment, the company negotiates the right to operate and earn a profit from the new system for a specified number of years, after which the ownership of the system reverts to the municipality. There are only a handful of examples of this arrangement: Moncton, New Brunswick; Cranston, Rhode Island; Franklin, Ohio; and Tampa, Florida.

In all three public-private partnership arrangements, the municipalities establish the guidelines for operations and maintenance, and in the case where private companies will be building new facilities, the municipalities also establish the design criteria. Private companies then submit competing bids for the contract, and the municipality selects the best bid.

Of course, in every arrangement, it's still up to the municipality to make sure that drinking water regulations are being met and that the terms of the contract are being fulfilled. Ultimately, it's the municipality that is responsible to its citizens for ensuring safe drinking water, so municipalities still spend time and money to oversee privatized water systems.

Privatization can work well in many instances, particularly in small communities that lack the expertise or resources to operate water systems effectively. A private company can take advantage of economies of scale by providing scientific, technical and business services from one central location to a number of small, scattered water systems. However, there are also success stories in bigger cities: Moncton, New Brunswick, for example, has seen a substantial improvement in the quality of its water since a private company was contracted to finance, design, build and operate its water plant.

Despite the increasing popularity of these arrangements, there are several basic arguments against privatization. First, the primary duty of a privatized company is to its shareholders, not to its customers. Because the corporate focus is on short-term profits, private companies are less likely to consider the long-term public interest in protecting water sources and conserving water supplies. Consumers can expect to pay more for their water after privatization. Although some of this price increase reflects the real cost of providing drinking water, some of it reflects the shareholders' demand for profit. There is also the risk that customers who can't pay their bills will lose their water. In Detroit, where water is supplied by DTE Energy, more than 40,000 people had their water cut off for non-payment between July 1, 2001 and June 30, 2002.

Private companies are also much less open to public scrutiny, and there is less public access to information. Consumers in many North American municipalities have found it extremely difficult and frustrating to have questions answered or complaints dealt with under these circumstances. Additionally, split jurisdiction between a private company and the municipality can prove frustrating for consumers. The company may refer complaints and concerns to the municipality, while the municipality feels the same issue should be dealt with by the company.

Furthermore, water suppliers essentially have a monopoly, so the common rationale for privatization — opening up a sector to market forces — doesn't really hold true in this situation. Although the initial bidding process is competitive, once a private company has won a contract, it doesn't have to worry about competition from other suppliers. Consumers can't shop around for better water or better service if they are unhappy, except by moving to another region.

While private companies generally deliver on promises to operate water systems more efficiently — saving money at an average of 10-25 percent a year — this efficiency comes at a cost. Privately run water suppliers generally have fewer staff and invest less money in maintenance and upkeep, which may threaten the long-term health of a water system and its capacity to deal with emergencies.

Finally, the private water sector has proven quite volatile in recent years. There have been numerous takeovers, the threat of a few bankruptcies, and the high-profile failure of Enron-owned Azurix. Consider the case of Hamilton, Ontario: since it contracted out its water services in 1995, ownership has changed hands four times. The initial contractor, Philip Utilities Management Corporation, was sold to Azurix Corporation when its parent company teetered on the verge of bankruptcy. The pattern repeated itself when Enron was plunged into corporate scandal — it sold Azurix to American Water Works. Then in 2001, American Water Works was taken over by water giant RWE. One Hamilton city councilor describes the situation as a "revolving door." It's difficult for municipalities and citizens to obtain information, demand accountability or develop relationships with managers when the water system is owned by so many different companies in rapid succession.

Several American municipalities that privatized their water systems are choosing not to renew contracts. Some are actually buying back their systems or cancelling contracts. For example, Lee County, Florida outsourced the operation and maintenance of its water and sewer system in 1995 to ST Environmental Services. Although the company did cut costs and increase efficiency, an audit five years later revealed problems with treatment procedures, inadequate maintenance, poor handling of hazardous materials and long delays in customer billing. In 2000, the county's Board of Commissioners voted to return water and sewer operations to public control. The county utilities director subsequently determined that ST's failure to maintain infrastructure properly would cost US$8 million (Can$12 million) to rectify.

Perhaps the greatest blow to privatization proponents in North America has been Atlanta's high-profile decision in January 2003 to cancel its contract with United Water. Atlanta was supposed to be the "jewel in the crown" of private water suppliers in North America — proof that the private sector could run the water system of a major metropolitan area effectively and efficiently while reaping significant savings for the city. The bidding process was fiercely competitive. United Water was the ultimate winner, walking away with a 20-year, $20.8 billion (Can$33.2 billion) annual contract. Four years later, the results have been disappointing. The city cited unrepaired leaks, interruptions in service, boil water orders, failure to collect overdue water bills, and only half the savings Untied Water had promised to deliver.

In its defense, United Water claimed Atlanta had not revealed the extent of its infrastructure woes — century-old pipes suffering from years of neglect. The company was forced to invest much more than it had forecast on maintenance and repairs, and as a result it was losing $10 million a year. In the end, both parties maintained that the termination of the contract was amicable. According to Atlanta's chief policy officer Greg Giornelli: "The residents of Atlanta cannot get good water under the contract, and United Water cannot make money under the contract." The city has created a 346-person department to run the new public system.

BULK WATER EXPORTS

Exporting water in bulk is another idea that is making a lot of people nervous. The concept is quite straightforward. Take Canada as an example — it's home to nine percent of the world's freshwater. Piping some of this plentiful resource to thirsty markets seems like a good financial proposition, and there's certainly no shortage of interest in export schemes — and no shortage of controversy either.

Nova Group, a company in Sault Ste. Marie, Ontario, had plans to ship 793 million gallons (3 billion liters) of water a year from Lake Superior under a five-year permit from the provincial government, but the permit was subsequently revoked, and Ontario has now endorsed a national accord prohibiting bulk water exports.

In British Columbia, the provincial government granted an export permit to the California-based Sun Belt Water, who proposed to export British Columbian water to the California market. British Columbia turned around and placed a moratorium on bulk water exports, possibly to avoid setting dangerous precedents under NAFTA, and Sun Belt sued for US$10.5 billion (Can$14.3 billion) under NAFTA's Chapter 11.

Most recently, Newfoundland is considering a plan to take 16 billion gallons (60 billion liters) of water each year from Gisborne Lake and ship it in bottles and bulk container ships across North America. To date, no decision has been made.

What's at stake is control over a natural resource that is both vital to human survival and an essential component of natural ecosystems. Environmentalists are concerned about the damage these schemes could mean for local watersheds: removing large volumes of water will affect the natural cycles, destroy wildlife habitat and fundamentally change landscapes. Other citizens are worried that Canada would be selling off a vital natural resource simply to feed the golf courses and swimming pools of the United States.

The debates don't stop there. Not only are there conflicts between those who want to export water and those who want to keep it where it is, it's not clear whether Canada has the authority to ban exports even if it wanted to, because of various trade agreements. Furthermore, if a ban were legal, it's something that would require joint agreement from the provinces and the federal government — not easy to achieve!

The legalities of such bans are complex and ambiguous, but the key points seem to be as follows:

Water in its "natural state" — in lakes, rivers and aquifers — probably isn't considered a "tradable good" under the North American Free Trade Agreement (NAFTA) or the General Agreement on Tariffs and Trade (GATT), although this isn't entirely clear. What's quite clear is that as soon as any company in Canada begins exporting water in bulk, water becomes a tradable good. And once that tap is turned on, trade agreements mean that it stays on.

Article 315(1) of NAFTA — known as the "proportional sharing" provision — means that if Canada chooses to export any water, the US is then entitled to a proportional share of Canada's water resources. This essentially means that once the US is granted a certain proportion of exported goods, that proportion can never be reduced, regardless of the need in Canada. Under Chapter 11 of NAFTA, once the government starts treating water as a tradable good, foreign investors will have the same rights to water resources as Canadian companies. And since natural resources, including water, fall largely under provincial jurisdiction, if any province decides to allow bulk water exports, all Canadian water will be up for grabs.

At the moment, no one has opened this Pandora's box. What will happen in the future, however, is anybody's guess.

CHAPTER EIGHT

The Bottom Line

WE HAVE COME A LONG WAY since the days of widespread cholera and typhoid caused by contaminated water supplies, but the fact remains that water treatment is not perfect. Human error happens from time to time, and treatment plants occasionally break down. As a result, disease outbreaks still occur each year, thanks to inadequate treatment or problems in the distribution system. On top of that, standard processes don't remove all the industrial and agricultural contaminants that make their way into water supplies. Many of these contaminants are newly created chemicals, and the combined effects of most of them are unknown. Some of them can be removed by advanced (and expensive) processes, but others cannot.

As we've seen, water treatment can also add undesirable substances into drinking water, such as bromates or THMs (see Chapter Three: *How Is It Treated?*). Finally, contamination can occur after treatment, such as when lead leaches out of water mains (see *Drinking Water Contamination*, Chapter Two).

ARE THE REGULATIONS STRICT ENOUGH?

US regulations are among the strictest in the world, although this shouldn't be a reason for complacency. There are a number of contaminants that have not yet been regulated but show up frequently in drinking water, such as the fuel additive MTBE. There is also the concern that consumers may be exposed to significant "spikes" of contaminants in their drinking water that may threaten health, even though the average level on the contaminant meets federal regulations. Pesticides are a good illustration of this — in the spring, when pesticides are commonly applied, the levels in surface water can increase dramatically, although the annual average may meet regulations.

In Canada, there's a lot of room for improvement. Although the federal government has issued a set of drinking water quality guidelines, developed in consultation with the provinces, they are not binding unless a province chooses to adopt them as legally enforceable regulations. (See *Chapter Seven: How Is It Managed?*) Some provinces actually go beyond the Canadian Guidelines. Alberta has enacted stricter microbial standards, while Quebec boasts the highest drinking water standards in Canada, with standards in place for 77 substances. Alberta and Quebec are the only provinces to demand filtration and disinfection for all surface water sources. Also, not all provinces require water system operators to be certified, and not all provinces require water testing to be done in accredited labs. These are all areas where regulations could be stricter. The regulations on how frequently suppliers must test for different contaminants vary from province to province, and could be improved significantly in many cases.

In most jurisdictions in North America, the regulations governing small water suppliers are not as strict as those for larger systems, recognizing the fact that smaller suppliers often lack the financial resources or the expertise to upgrade the treatment that they provide (see *Special Challenges for Small Systems,* Chapter Three). Alberta is the exception — all waterworks in the province must meet the same design performance standards, regardless of size.

It's important to recognize that the fact that water meets standards does not always mean there aren't harmful levels of contaminants in it. Setting maximum acceptable levels for many contaminants is an extremely difficult business. Ideally, standards should be based on documented health effects of long-term exposure, but as discussed earlier, the data is often insufficient or contradictory.

It is also impossible to set standards for all the different chemicals that might get into water supplies. New chemicals are being developed at an astounding rate. There are now approximately 70,000 chemicals in use, with 500 to 1,000 added each year. We don't know what the long-term effects of many of these chemicals are, nor do we know what happens when they combine with each other.

Essentially, it just isn't feasible or economic to remove all risk from a drinking water system. The goal of water suppliers and regulators is to reduce the risk to a reasonable level. So despite all the regulations and tests, water suppliers cannot necessarily guarantee safe water.

Finally, strict regulations are only part of the solution — they must be coupled with effective monitoring and enforcement to make them meaningful. The problem is, many jurisdictions do not have the money or staff to make sure that happens. In well over 90 percent of cases where water suppliers violated US standards, neither the state nor the EPA took enforcement action.

ARE WATER SUPPLIERS MEETING REGULATIONS?

Are water suppliers meeting regulations? The short answer is no, not always. The most recent statistics from the US reveal that in 2000, five percent of Americans were served by community water suppliers that violated maximum contaminant levels, and six percent were served by community water suppliers that violated treatment techniques — both types of violations that can potentially endanger human health. In addition, 13 percent of Americans were served by community water suppliers that violated monitoring and reporting requirements. Very small systems were 50 percent more likely to violate standards than larger systems, although this performance has been improving.

In a number of cases, water suppliers have been granted temporary exemptions from new regulations to give them time to upgrade their systems. Even so, many suppliers take longer than allowed to implement new monitoring schedules or upgrade their treatment processes. The EPA has seen clear trends in increased violations for several years after new regulations come into effect.

In some cases, the EPA has taken water suppliers to court for chronic violations of state and federal drinking water standards. Many recent lawsuits have accused suppliers of failing to filter surface water — a process required by the EPA to reduce the risk of *Cryptosporidium* contamination.

In terms of waterborne disease outbreaks, the US Centers for Disease Control reported a total of 39 in 1999—2000. However, this is probably a massive underestimate of the real number. The vast majority of waterborne disease incidents are never identified or reported, or are never traced to contaminated drinking water. There are a number of reasons for this. Many people do not seek medical attention for tummy upsets, and even when they do, physicians usually cannot attribute a few cases of gastrointestinal illness to a particular source.

Based on the statistics that do exist, parasites are the most common cause of waterborne disease outbreaks in the US, accounting for roughly one third of outbreaks. In another one-third of cases, the contaminant was never determined. The remaining one third were traced to a variety of causes, including bacteria, viruses and chemical contaminants.

In Canada, it's difficult to look at national trends because the provinces have different water quality standards, different monitoring requirements, and different approaches to data collection. Health Canada has reported more than 160 waterborne disease outbreaks between 1974 and 1996, affecting about 8,000 people, but this is likely just the tip of the iceberg.

British Columbia has the dubious distinction of having the highest rate of intestinal illness in the country, due in part to waterborne diseases. Between 1980 and 2000, there were 39 confirmed waterborne outbreaks caused by micro-organisms like *Giardia, Cryptosporidium, Toxoplasma,* and *Campylobacter.* Many of these outbreaks were traced to water system failures or the absence of adequate treatment. In August 2001, ten percent of the water systems in the province were under boil-water advisories because the water did not meet minimum treatment standards or because coliform bacteria were detected in water samples.

In Newfoundland, there were 302 boil-water advisories in effect in 2002, affecting approximately 66,500 people. The most common reasons for issuing the advisories were lack of disinfection and problems with residual chlorine levels. Other water quality problems included turbidity, THMs and pH.

Fifteen percent of First Nation water systems fail to meet health-related guidelines, thanks to inadequate infrastructure and lack of trained personnel to operate and maintain the systems.

WHAT IF YOU'RE PARTICULARLY VULNERABLE

If your immune system is compromised for any reason, your tap water may not be safe to drink. Cancer patients undergoing chemotherapy, people with HIV/AIDS, organ transplant recipients, and some elderly people should consult with a doctor to discuss boiling drinking water, filtering it, or relying on bottled water to avoid microbial contaminants like *E. coli, Giardia* and *Cryptosporidium.* (See Chapter Two: *What's In It?* for specific recommendations for each contaminant.)

Infants are also more vulnerable to contaminants, but boiling their water isn't always advisable. Although it will kill any pathogens present, it will also concentrate contaminants such as lead or nitrate. Pregnant women should be concerned about lead and nitrates, which can harm the fetus. Filtering your water with a filter system certified to meet ANSI/NSF standards for the contaminants in question is a good option.

CHAPTER NINE

Making Do With Less

IS THERE ENOUGH?

NORTH AMERICA IS RICH in water resources. Canada especially is blessed with abundant freshwater. Unfortunately, we often take these resources for granted and squander them, hosing down the driveway, or flushing cigarette butts down the toilet. Attitudes are beginning to change, but there's a long way to go.

Water Consumption

How much water do we use? Per-capita consumption in Canada has been gradually decreasing since an all-time peak in 1989. In 1999, Canadians used 169 gallons (638 liters) per day. In the US, per-capita consumption was about 185 gallons (700 liters) per day in 1990, with highest consumption levels in the western states, thanks to their arid climate.

Not all of this water gets used in the home: per-capita consumption is calculated by taking all the water used across the country and then dividing it by the number of inhabitants. It includes water used in agriculture, industry and commercial sectors, and it can fluctuate quite a bit, depending on the economy and the weather.

Power generation accounts for 58 percent of water use in Canada and 39 percent in the United States (see Figure 9.1, page 78). This is considered "non consumptive use" — the water that turns the hydro turbines isn't taken away or polluted in the process, so it's available for other uses downstream. Agriculture is the single biggest user of water in the US, where it accounts for 42 percent of total water withdrawals. In both countries, municipal use is one of the smallest slivers of the pie.

Figure 9.1 Water Use by Sector

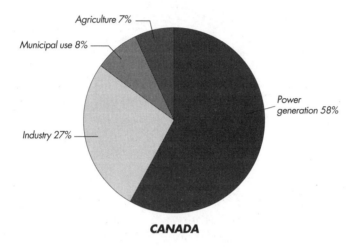

CANADA

- Agriculture 7%
- Municipal use 8%
- Industry 27%
- Power generation 58%

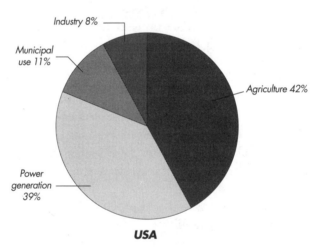

USA

- Industry 8%
- Municipal use 11%
- Power generation 39%
- Agriculture 42%

(Source: UNEP, GEO-2000)

If we focus strictly on household water use, Canadians use 90 gallons (343 liters) per person per day. Americans consume more — over 148 gallons (560 liters) per person per day. In fact, Americans are the greatest consumers of water in the world, with Canadians placing a close second. Clearly there's room for improvement here.

Up Against the Limit?

Although North America has an abundance of freshwater compared to many other parts of the world, that doesn't mean that water shortages don't occur. Shortages are an ongoing issue in many areas, particularly in California and the southwestern United States. Under the Great Plains, the water table of the massive Ogallala aquifer is steadily dropping: throughout the 1980s, the water table dropped by an average of 6 inches (15 centimeters) a year; in 1994 it dropped 2 feet (60 centimeters); and in 1995 it fell almost 3 feet (1 meter). Tucson, Arizona has been forced to increase the depth of its wells from 500 feet (150 meters) to 1,500 feet (450 meters) to ensure a steady supply of water.

What Makes a Drought?
It doesn't take a big change in rainfall patterns to cause a drought. Most of the water supply for the summer and autumn is collected in the winter, when much less water is lost through evaporation and transpiration. A drop of 20 percent in winter rainfall can mean the reservoirs don't fill up, and that almost certainly leads to water restrictions in the summer. Of course, any time there is a prolonged dry period, there is a risk of drought, especially if surface water is the main source of drinking water, but dry winters are particularly dangerous.

Across the continent, water restrictions are a regular occurrence each summer, and droughts have made headlines in recent years. In 1999, 26 percent of Canadian municipalities reported problems with water availability in the previous 5 years. During the summer of 2002, almost half of the United States was hit with moderate to extreme drought. The 2001 drought caused millions of dollars of agricultural losses in Canada, and 42 prairie communities suffered under severe water shortages. Unfortunately, hot, dry summers are expected to become more and more common in North America, thanks to the impact of climate change.

Water shortages are forcing many communities to look at ways of reducing water consumption. It also makes sense from an economic point of view: consuming less water keeps treatment costs down. And there is another compelling reason to cut water consumption: water quantity is closely linked to water quality. If surface water levels drop, the pollution in the lake or river becomes more concentrated, and it's going to take more treatment to produce drinking-quality water.

Consuming too much groundwater can cause a special problem in coastal areas such as Prince Edward Island and Florida: saltwater

Technological Silver Bullets

As water becomes increasingly scarce in many parts of the world, the search for a technological fix is on. The best-developed of these is desalination, often hailed as the "silver bullet" that will solve our water woes. The premise is simple: there are vast amounts of water in the world's oceans, but it's saltwater. All we need to do is extract the salt from the seawater to produce an almost unending supply of freshwater. What's holding us back, in a nutshell, is money. While the technology has been in place for decades, the cost of desalination continues to be prohibitively expensive. The energy required to remove the salt ions from seawater doesn't come cheaply. For the foreseeable future, desalination doesn't make sense except in areas with large amounts of energy to spare and extremely limited sources of water – areas like the Persian Gulf. In North America, water conservation is the much more sensible way to go.

intrusion. If too much groundwater is taken from these aquifers, salt water will flow in to take its place, making the water unfit to drink. Saline intrusion is difficult and very expensive to reverse — in most cases, the wells are simply abandoned and new sources of water have to be found.

REDUCING WATER CONSUMPTION IN THE HOME

As a simple experiment one day, why not calculate how much water you use? Every time you turn on a tap or use a water-consuming appliance, measure the quantity, and at the end of the day add it up. You'll be able to see for yourself where you can best make savings. If your water supply is metered, it's even easier — just read your meter at the beginning and end of the day, and then calculate the difference.

Most households can easily cut water consumption by 25-30 percent by fixing leaks, reducing the amount of water used for flushing toilets, and installing water-efficient showerheads and faucet aerators.

In the Bathroom

Toilet flushing is the single biggest use of water in the home, accounting for roughly one third of household water use (see Figure 9.2, page 81). The majority of toilets in North America use 5 gallons (19 liters) per flush. Suppose you flush your toilet just six times a day: that adds up to more than 30 gallons (110 liters) per person simply to flush away waste.

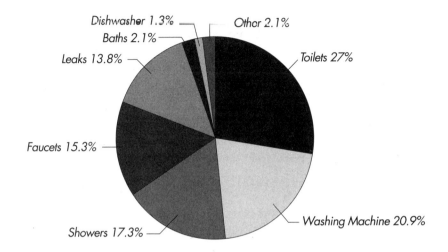

Figure 9.2 Typical Indoor Water Use in a Single-family American Home

Toilet flushing accounts for almost one third of all indoor water use in a typical American household. (Source: AWWA, 1999)

Fortunately, there are many ways to reduce this number. If you can afford to replace your toilet, there are many water-saving models available. In 1992, the US set a national manufacturing standard of 1.6 gallons (6 liters) per flush. The EPA estimates that the water saved by using these high-efficiency toilets in all new construction and through normal rates of replacement in older buildings will save over 7.6 billion gallons (29 billion liters) per day by 2020. Likewise, Ontario insists on 1.6 gallon (6 liter) toilets in all new buildings.

Low-flow models can be quite efficient — a good toilet depends more on the velocity of water flushing the bowl than the volume of water used, and many new models use a variety of technologies to create a more forceful flush. Dual flush models offer the option of a short flush or a long flush.

Low-flow toilets have earned a bad reputation in the past — older models were well known for problems with clogging and the need for double flushing. Don't let that hold you back! Recent models are much improved, and several different surveys have found that owners are happy with the performance of this "next generation" of low-flow models. A 1999 customer satisfaction survey published by the Metropolitan Water District of Southern California found that more than half the respondents said their new 1.6-gallon-flush toilet clogged less than the 3.5 gallon model it replaced.

Having said that, you can probably expect to double flush your low-flow toilet from time to time, and some models are much better than others, so it pays to shop around. There are three main types of low-flow toilets to choose from. Gravity-flush toilets are similar to the old-fashioned, 5 gallon (19 liter) per flush model. They're going to be your cheapest option, both to buy and maintain, but they don't perform as well as the other varieties.

Pressure-assisted toilets use household water pressure to compress air, which in turn blasts water into the bowl and flushes your waste away. This is a very effective system, so long as your water pressure is at least 25 psi. The biggest drawbacks are price — pressure-assisted toilets are expensive to buy and maintain — and the noise of the flush.

The third (and least common) variety is the vacuum-assisted toilet. It contains vacuum chambers in the tank to suck waste down the trap. These models are reasonably priced and not expensive to maintain, and they tend to perform well. They are also much quieter than the pressure-assisted varieties.

An excellent place to start researching low-flow toilets is the October 2002 issue of *Consumer Reports,* which features an assessment of 28 different models. Canadians in the market for a low-flow toilet should be aware that CSA certification does not guarantee good performance. A "field testing" program in Toronto found problems with a number of toilets that received a CSA stamp of approval, so the *Consumer Reports* ratings are a better guide.

Many cities are actively encouraging residents to replace their water-guzzling toilets with more efficient models. Santa Monica, California ran a very successful program offering residents rebates on high-efficiency toilets. As a result, Santa Monica has reduced total water demand by 15 percent. Galeta, California did even better — a toilet and showerhead rebate program, coupled with a public education program and increased water rates, resulted in a 50 percent drop in per-capita residential water use.

Compost toilets are the ultimate in water conserving models — they use no water at all. Small electric models exist which are only slightly larger than a normal toilet and use the same amount of electricity as a fridge. The deposits are heated to evaporate moisture and a fan extracts smells in some models. They fit in a conventional bathroom, and the residue they produce can be composted and used on the flower beds. The environmental benefits are offset to a certain degree, however, by the fact that these toilets use electricity.

Larger composting toilets are available which do not use any electricity, but which do take up more space. If you have such space available, these models are far more beneficial, with no environmental drawbacks. Composting toilets may seem radical, but they've been used very successfully at the C.K. Choi Institute at the University of British Columbia for the past seven years. The building is a 30,000 square foot office complex, and it relies on 10 flushless toilets and a number of flushless urinals to serve the needs of all its occupants.

If you can't afford to replace your toilet with a low-flow or composting model, you can reduce the volume of the flush by installing toilet dams or placing a plastic bottle filled with water in the tank (see Figure 9.3 below). Another very simple water-saving technique is not to flush the toilet after every use.

Figure 9.3 Reducing Your Toilet Flush

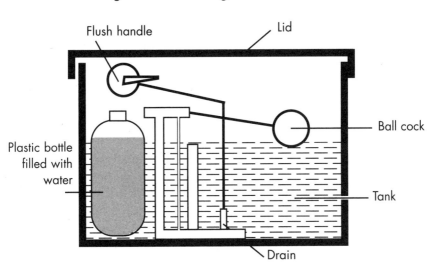

Placing a plastic bottle filled with water in the toilet tank will reduce the amount of water used for each flush. Be careful the bottle does not interfere with the ball cock mechanism. You should also be careful not to reduce your flow too much — depending on the design of your toilet bowl, you may need as much as two gallons (eight liters) to flush effectively.

Bathing and showering account for about 20 percent of household water use, so a simple way to economize is to use less water in a bath or take shorter showers. Showers generally use less water than bathing, but that clearly depends on how long you spend under the showerhead. Install low-flow showerheads to maximize water coverage and minimize water volume — you'll be using half the water you used to, without sacrificing any enjoyment.

In the Laundry Room

Washing machines are the second-largest water user in your home. Put only full loads in your washing machine. If you're in the market for a new washer, look for economy features such as half load capability or reduced water consumption. Front loading washers are a great choice, using 38 percent less water than top loading models. Choose a model with Energy Star mark, an international symbol which means the appliance meets energy-efficiency standards.

In the Kitchen

If you wash dishes by hand, use one bowl for washing and one for rinsing. Bowls hold less water than it takes to fill the sink. If you use a dishwasher, make sure it's full before you run it. Don't waste water by rinsing your dishes before you put them in the dishwasher, unless they're going to sit there for a while before you run the machine. *Consumer Reports* tests have shown that pre-rinsing does not improve your dishwasher's performance.

Don't run your tap more than necessary. Keep a jug of drinking water in the fridge, so you avoid running the tap for long periods waiting for cold water. Conversely, insulate hot water pipes to prevent running the tap for long periods waiting for hot water.

When it comes to vegetables, clean them in a bowl of water, not under a running tap. Compost your peelings instead of using a waste disposal unit. After you've cooked your veggies, consider saving your cooking water to use as a stock or a base for soups; it can be kept for several days in the fridge, or for longer in the freezer.

In General

You can reduce your water use by as much as 14 percent simply by stopping leaks. Check for dripping taps and worn washers; mend or replace them right away. A leak of one drop a second adds up to an astounding 2,600 gallons (10,000 liters) a year.

A leaking toilet wastes even more — up to 53,000 gallons (200,000 liters) a year. You can find out whether your toilet leaks by putting a few drops of food coloring in the tank. Don't flush! If any food coloring begins to show up in the bowl, you have a leak. The most common causes of toilet leaks are flush valves or flapper valves that aren't sitting properly on the valve seat, corroded valve seats, or flush valve wires that are misaligned. These are all fairly simple to fix. If your toilet is leaking from the base, however, it's time to call in a plumber.

Don't forget to install spray taps or aerators on all your faucets. You'll use a smaller volume of water to achieve the same results as your old steady-flow taps.

Outside

Water consumption doesn't stop at your front door: 50-70 percent of household water use in the US is outdoors. There are many ways to trim this figure.

First of all, water smarter. To save water and to give your plants the maximum benefit, water out of direct sunlight — in the evening for example. This will cut down on water loss due to evaporation. Avoid sprinklers, which use water indiscriminately. Try to target the water precisely where it is most needed. Infrequent, deep watering is better for your plants than frequent, shallow watering.

Tampa, Florida cut 25 percent of its water use through its "Sensible Sprinkling" program. Through door-to-door visits, city staff evaluated homeowners' watering practices and suggested ways to reduce outdoor water consumption. The city promoted outdoor water efficiency by liming outdoor irrigation to one day a week and prohibiting watering between 8 a.m. and 6 p.m. It also created landscape codes that limit the amount of irrigated turfgrass in commercial landscapes to 30 percent.

Whenever possible, grow your ornamental plants and veggies in beds rather than containers, which dry out more quickly. Grow drought-resistant plants which don't need so much water — native plants are a good choice, since they're naturally adapted to your local climate. And consider reducing the size of your lawn: grass tends to be a big water-guzzler. El Paso, Texas saved 1 billion gallons (1.6 billion liters) of water in 2001 thanks in part to a program that reimbursed residents who replaced their lawns with water-efficient landscaping.

When it comes to lawns, longer is better. Set your lawn mower blades so that the grass is at least 3 inches (7.5 centimeters) high after you've cut it. Longer grass reduces the amount of water that evaporates, so you won't need to water as much. Longer blades of grass also encourage the growth of deeper roots, making your lawn more drought-resistant.

Rainwater collected through your eavestroughs is ideal for watering the garden, since it's not chlorinated. It's also good for washing your car. It is quite simple to collect rainwater in a barrel placed under your downspout, and many municipalities sell barrels at a discount. Keep the barrel covered to prevent algae and insects from breeding in your store of rainwater, or add a few drops of vegetable oil to create a surface barrier. For convenience, you can install a tap at the base of the barrel and hook up the hose whenever you're ready to do some watering. (For more elaborate rainwater systems, see *Obtaining an Independent Supply,* Chapter Twelve.)

Don't use a hose to wash your car. It is possible to do the job with only two buckets of water — one of soapy water and one of rinsing water. And if the sidewalk or driveway needs cleaning, choose a broom, not a hose. Finally, if you have a pool, keep it covered when it's not in use, to avoid evaporation.

REUSING WASTEWATER

A final strategy for reducing water consumption is to reuse wastewater for things like flushing toilets or watering the garden, where it's not necessary to have drinking-quality water.

Obviously, reusing "blackwater" — the highly polluted wastewater from toilets — is out of the question, but what about "greywater" from everything else: sinks, bathtubs, washing machines, etc.? Greywater is relatively clean, and it rarely contains disease-causing organisms, unless you're washing dirty diapers. Greywater reuse is not common in Canada, but it is more widely practiced in the US, particularly in California.

On the face of it, it makes sense to reuse greywater for watering the garden or flushing the toilet, for example. In practice, it requires a fair bit of effort and may even be prohibited during droughts (because your wastewater may be needed as a source of water downstream). Codes, standards and regulations may also make water recycling difficult, so check with your local municipality before you plunge in. Greywater does contain significant quantities of grease, food waste, detergent and soap residues, hair, and other dirt, so you need to be a little discriminating in how you use it. If you decide it's worth doing, keep the following points in mind:

- Put a mesh filter on your plug hole — this will screen out the biggest particles.
- Use biodegradable cleaning products without additives if you plan to use your greywater for irrigation; avoid boron and bleach.
- Don't try to store greywater — what you'll end up with is an unhealthy, smelly bacterial soup.

How much can you reuse? The average household produces approximately 26 gallons (100 liters) of greywater per person per day. If you plan to use it for irrigation — the most common form of reuse — figure that one square foot of soil can deal with an average 3 gallons per week (10 liters per square meter pre week). You'll be able to add more in the summer, but less in the winter. Scale your greywater reuse to suit your irrigation needs, and use the cleanest source first. This tends to be bath and shower water, followed by water from the bathroom sink, washing machine, utility sink, dishwasher and kitchen sink. If the volume of water from your bathtub is enough to water your garden, there is no need to worry about reusing any from the kitchen sink.

When it comes to irrigation, don't just throw your greywater on the garden. Unregulated dumping will mess up the soil structure, damage root hairs, and make clays more sticky and unworkable. And because soaps and detergents are rich in sodium, greywater can increase soil alkalinity. Some plants like azaleas, conifers and rhododendrons prefer acidic soil, so avoid watering them with greywater. The best approach is to rotate between different patches of soil — don't irrigate one spot exclusively or continuously. It's also a good idea to alternate between greywater and clean water. Finally, because there is a slight possibility that pathogens could be present, it is safest not to use greywater on food crops.

CHAPTER TEN

If You're Not Happy

FINDING OUT WHAT'S IN IT

AS A CONCERNED CITIZEN, you should know what's in the water you drink. If you live in the United States, finding out is quite straightforward. All public water suppliers are required to produce an annual report on the quality of their water. Generally, these Consumer Confidence Reports are mailed to customers and many suppliers also post these on the Internet. In the case of some smaller suppliers, you may need to call and ask for a copy to be mailed to you.

Deciphering the report may not be quite so simple — a survey of 1999 Consumer Confidence Reports across the country conducted by the Campaign for Safe and Affordable Drinking Water found that although some reports were very clear and easy to interpret, others were confusing or misleading. Hopefully this situation will improve now that the EPA is providing water suppliers with examples of how to present their data clearly. If you have any difficulty making sense of the information in your report, call your water supplier and ask for some clarification, or consult the brochure put out by the Campaign for Safe and Affordable Drinking Water entitled *Making Sense Out of Drinking Water "Right to Know" Reports* (see Resources).

If your report contains a blanket statement that your water is totally safe, don't stop reading. Check to see whether your water supplier violated any EPA standards — either Maximum Contaminant Levels or Treatment Techniques. These indicate potential health hazards.

For each contaminant, your report should list the Maximum Contaminant Level Goal or MCGL (an unenforceable health-based goal), the legally binding Maximum Contaminant Level or MCL, and whether a violation occurred. It will also tell you how much of the contaminant was detected on average, and what the range of detected levels was.

In some cases you may see that the top end of the range was over the EPA standard but no violation occurred. That's because for many contaminants, it's the average of many samples that determines whether or not a violation has occurred. Unfortunately, that means that in some parts of the distribution systems, THMs may be above the EPA standard. Or it may mean that you're exposed to high levels of pesticides or nitrates at certain times of year, although your average annual exposure is within EPA limits.

Your report should also tell you how frequently the water was tested for each contaminant, and the likely source of contamination. If any violations occurred, the report should explain the potential health effects of that contaminant and what the supplier did to address the problem.

Keep in mind that some important contaminants like radon, some infectious parasites, and many types of pesticides are not covered by EPA regulations, so you won't find any information about them in your Consumer Confidence Report.

In Canada, you'll probably need to do more digging to determine exactly what you're drinking. Although some cities and provinces provide reports on your drinking water quality, many do not. Begin by contacting your water supplier, and if that doesn't work, try your provincial Ministry of Health or Ministry of the Environment (see Resources). Some provinces have databases of community water quality that you can access on line.

You can also have your tap water tested (see Chapter Twelve: *If You're Not on the Mains*). Home test kits are available, but these do not tend to be as comprehensive or accurate as an analysis by a registered laboratory. Keep in mind that testing just once will give you a snapshot of your water quality, but it won't tell the whole story. For that, you'd need to test your water on a regular basis, and that won't come cheap.

If there are problems with your water quality — high levels of heavy metals, for instance, or unpleasant tastes due to chlorine residuals — you might consider filtering your tap water. (See Chapter Twelve: *Alternatives to Tap Water*.)

MAKING A COMPLAINT

If you're not happy with the quality of your tap water, the best place to start is by contacting your water supplier. In many cases this will be your local municipality; in some cases it is a private company or a provincial agency. You should be able to find a phone number on your water bill, in your telephone directory, or by contacting your town hall. Ask why the problem you're concerned about is occurring, and how the water supplier intends to address it.

If you're not satisfied with the response you get from your water supplier, your next step is to contact your state water supplier or the relevant provincial ministry. In most Canadian provinces there isn't a single agency with overarching responsibility for drinking water issues; instead jurisdiction is usually shared between at least two ministries, and sometimes four or five. In this case, your best bet is to try the public health department of your provincial Ministry of Health or the water division of the Ministry of the Environment. In some Canadian provinces, the Ombudsman can be consulted for mediation assistance.

In the US, you also have recourse to the federal Environmental Protection Agency. You can call the EPA Safe Drinking Water Hotline or contact your regional EPA office (see Resources). As a last resort, American citizens have the right to sue jurisdictions that fail to meet water standards.

In Chesapeake, Virginia, more than 200 women launched lawsuits against the town, claiming that high levels of THMs in the municipal water supply caused their miscarriages. Furthermore, they allege that water supplier falsified results to show lower THM levels — something the city was fully aware of. The lawyer for the city agrees some THM reading were over the EPA's limit, but maintains that the seasonal averages were always below the Maximum Contaminant Level. The first test case went to court in February 2003.

In other cases, citizens have sued companies that polluted their water supply. There are a number of class action lawsuits currently working their way through the US courts against oil manufacturers and distributors, claiming the gasoline additive MTBE has contaminated public and private wells. MTBE is highly water soluble, moving quickly through the soil and into the groundwater below.

GETTING INVOLVED

If you're really concerned about your water, consider working to create long-term change. There are lots of ways to get involved in protecting and improving drinking water quality. Many communities have some kind of sourcewater protection program, administered by the municipality, state or province, or by a nonprofit group. Chances are, there will be no shortage of volunteer opportunities to conduct contaminant inventories, public education campaigns or watershed restoration projects.

Make your voice heard at the decision-making level. Many jurisdictions have committees that set policies and regulations on drinking water, and there's often room at the table for a member of the public, or opportunities for public input. The US EPA invites public review and comment on its drinking water regulation, and there's opportunity for input at the state level as well.

Finally, you can lobby for change by getting involved with an environmental or public interest group working on water issues. Some of the most prominent of these organizations are listed in the Resources section at the back of this book.

CHAPTER ELEVEN

Alternatives to Tap Water

MANY PEOPLE BUY BOTTLED WATER or use water filters because they believe it is healthier than drinking tap water. Bottled and filtered waters are certainly more expensive than tap water and they can taste better, but they may not be significantly better for you.

Before you reach for your wallet, do a little sleuthing. First of all, decide what your problem is: are you concerned about taste, or is it contaminants that have you worried? If contaminants are the issue, find out what's currently in your tap water so you know exactly what it is you want to remove. If you decide to go the home filtering route, look for ANSI/NSF certification and choose a system certified to remove the contaminants you're concerned about. If you prefer bottled water, read the label carefully: it may not be any more pure than your tap water. Finally, if the only thing you're concerned about is a chlorine taste in your water, the cheap solution is to leave an uncapped jug of water in your fridge for drinking purposes. You should find the chlorine taste is gone after the water has sat for 12-24 hours.

BOTTLED WATER

American consumers spend $4 billion a year on bottled water, with sales tripling in the past ten years. Growth has also been strong in Canada, with consumption of bottled water increasing ten percent per year through the 1990s. Americans knocked back almost 12 gallons (44 liters) per person in 1997, according to the bottled water industry, while Canadians drank a more modest 5.6 gallons (21.2 liters) each. That still leaves lots of room to grow — compare these numbers

> ### Categories of Bottled Water
>
> **Spring water** *must come from an underground source and must flow naturally to the earth's surface. Carbon dioxide can be added to make it sparkling.*
>
> **Mineral water,** *like spring water, must come from an underground source and must flow naturally to the earth's surface. In addition, it must contain a specified level of naturally occurring dissolved minerals. In the US the minimum level is 250 parts per million (ppm); in Canada it's 500 milligrams per liter. Dissolved minerals usually include calcium, magnesium, sodium, potassium, silica and bicarbonates. Be aware that the amount of minerals you'll get from mineral water is quite small compared to what you'll get from food, so don't count on bottled water as a significant source of essential minerals.*
>
> **Purified drinking water** *is bottled water that has been treated by distillation, deionization or reverse osmosis. The source does not need to be named, and it may even be municipal tap water, so don't be misled by pictures of pristine mountain glaciers or sparkling streams on the label. However, the label must indicate what kind of purification process the water has undergone.*
>
> **Artesian water** *is water from an artesian well. This means the pressure within the aquifer is high enough that it actually pushes water to the surface once a well is drilled.*
>
> **Well water** *is water from a normal well.*

with the 26.4 gallons (100 liters) that Germans drink each year, or the 35.6 gallons (135 liters) consumed by Italians.

Manufacturers credit this explosion in North American sales to consumer desire for pure, safe water, coupled with the increased availability of bottled water. Go into any convenience store, gas station, supermarket or food court, and chances are you'll be able to choose from a half-dozen different brands of water.

Whichever bottled water you choose, you're going to pay substantially more for it than you pay for tap water. So what's the difference?

In both Canada and the US, bottled water is regulated as a food product, not as drinking water. In Canada, Health Canada is the agency responsible; in the US, it's the Food and Drug Administration (FDA). Both countries define several different categories of bottled water (see *Categories of Bottled Water*, left), while certain types of bottled water are actually regulated as soft drinks, including soda water, tonic water, seltzer and any kind of flavored water.

According to FDA regulations, bottled water must be tested for microbiological contaminants on a weekly basis. For chemical contaminants, it's once a year, and for radioactive contaminants, it's once every four years. In theory, bottled water should be at least as strictly regulated as public tap water. The FDA must create bottled water regulations for any contaminant covered by the EPA's drinking water regulations, unless the contaminant isn't relevant to bottled water — like asbestos and acrylamide, for example, which are byproducts of the drinking water treatment and distribution process. If a bottled water fails to meet FDA regulations, the label must state that it contains "excessive" levels of the contaminant in question.

However, there is a very large loophole. In the US, water that is bottled and sold within the same state is exempted from FDA regulation. According to the Natural Resources Defense Council, this means 60-70 percent of bottled water sold in the United States. Instead, this water falls under state regulations — if such regulations exist. Approximately one in five states do not regulate bottled water.

For those bottled waters that are regulated by the FDA, the regulatory oversight may be quite lax. According to a report by the Natural Resources Defense Council, the FDA has admitted bottled water is a low priority for the agency. It has the equivalent of less than one staff person developing bottled water regulations, and less than one person monitoring compliance. If a product is recalled because it fails to comply with FDA regulations, it is generally left up to the bottler itself to voluntarily make the recall.

The picture is similar at the state level: in those states that do regulate bottled water, there is often the equivalent of less than one person dedicated to the job of developing and monitoring regulations.

Many bottled water manufacturers go above and beyond FDA requirements. Some choose to follow the International Bottled Water Association's Model Code, which calls for daily testing for total coliforms and other microbiological contaminants. Members of the IBWA must pass annual, unannounced plant inspections conducted by an independent third party. Check the IBWA website for a list of member manufacturers (see Resources).

In Canada, bottled water falls under the federal Food and Drugs Act which requires bottled water manufacturers to adhere to quality standards and good manufacturing practices, as well as labeling requirements. These regulations are legally enforceable across the country, unlike Canada's Drinking Water Quality Guidelines.

Health Canada is currently revising the regulations on bottled water to take into account new scientific knowledge — most of the existing regulations date back to 1973 — and to harmonize them with US laws, Quebec laws, and the Canadian guidelines for tap water. The new regulations will probably include stricter limits on some bacterial and chemical contaminants.

The Canadian Food Inspection Agency (CFIA) is responsible for inspecting bottled water manufacturers and for analyzing samples from both domestic and imported bottled waters. On average, they visit bottling plants every 12 to 18 months. Although there has never been a case of waterborne disease linked to bottled water, tests on bottled water quality aren't very reassuring. Between 1992 and 1997, just under one third of the bottles tested failed to meet Health Canada standards. It's useful to know that most of the bottled water that didn't pass was non-carbonated. Carbonation makes water more acidic, so bacteria have a harder time growing in it. In addition to federal CFIA inspections, bottled water is tested and monitored by some provincial and municipal agencies.

Canadian bottlers may also choose to become members of the Canadian Bottled Water Association (CBWA) and follow its Model Code. This involves passing annual, unannounced plant inspections conducted by an independent third party, including an annual water analysis at an independent laboratory for more than 150 possible contaminants. In addition, the bottlers themselves must regularly test for microbiological contaminants. Check the CBWA website for a list of their members (see Resources).

If you drink bottled water only occasionally, there's no need to be too concerned with reading labels. But if you consume it regularly, you should know what's in it. Most bottled waters give an analysis of their content on the label, and you can also contact the company and ask for more information. Look out for high levels of potassium — more than 12 ppm can put stress on your kidneys. Magnesium sulfate and sodium sulfate are laxatives — don't use high-sulfate water when you're making baby formula. Finally, you should avoid

The Great Perrier Benzene Scare

In February 1990, Perrier discovered their famous bottled mineral water was contaminated with benzene, a colorless solvent known to cause cancer. After much panic, it turned out that its source was not affected; the problem was traced to a faulty gas line filter at the bottling plant. It was a costly mistake: Perrier withdrew all its stock from store shelves and lost a significant part of its market share as a result of the incident. In the US and Canada alone, the recall added up to 70 million bottles.

waters with high sodium levels if you've been told to watch your sodium intake.

If you're using bottled water for baby formula, make sure you boil it first to kill any bacteria or other pathogens that might be in it. You should also contact the bottler and ask about nitrate levels, since infants are particularly vulnerable to this contaminant and boiling the water will concentrate any nitrates that are there to begin with. Don't use bottled water with high levels of sodium to make baby formula.

Don't forget to check the bottling date and best-before date. It's generally considered best to drink bottled water within two years of bottling. Once you've opened bottled water, keep it refrigerated to prevent bacterial growth, and consume it within three to four days. Drinking directly from the bottle will introduce bacteria — not a good idea if you're going to cap the bottle and then drink from it again at a later point or share it with a friend. If you're using a water cooler, it should be cleaned regularly and disinfected with bleach to prevent bacterial growth.

As far as taste is concerned, a 2000 *Consumer Reports* investigation found little difference between different brands. What did affect taste was the type of plastic used in the bottle: waters in clear PET plastic tasted better than those in cloudy, softer HDPE plastic often used in one-gallon containers. Furthermore, the analysis found that polycarbonate plastic (the strong rigid plastic used in five-gallon water-cooler jugs) can leave residues of bisphenol-A, which is both a carcinogen and an estrogen mimic.

Generally, bottled water tastes better than tap water and filtered water, but it's also the most expensive. Be aware that it may contain bacterial levels that are substantially higher than tap water, although these are not harmful types of bacteria. (Incidentally, it is not just for taste reasons that mineral waters are often served with a slice of lemon — lemon juice contains chemicals that are fatal to many bacteria.) And keep in mind that drinking bottled water doesn't mean you can ignore the quality of your tap water, since contaminants like radon will be absorbed into your body when you're showering or bathing.

Finally, you may want to consider the environmental costs associated with bottled water. Think about the enormous number of plastic bottles used by bottled water manufacturers every year — 1.5 million tons of plastic, according to a 2001 study by the World Wildlife Fund. Bulk water extraction is becoming a hot issue in many areas where large volumes of water are being extracted from aquifers and rivers by bottling companies, affecting local ecosystems. There's also the environmental costs of trucking it long distances, adding to our air pollution and climate change woes. (See Chapter Seven for a more in-depth discussion of bulk water exports.)

FILTER SYSTEMS

Filter systems may improve the taste of your tap water, and they can be quite effective at removing a number of contaminants. They are increasingly popular — roughly one in five North American households filter their water. The greatest drawback (aside from cost), is that the consumer becomes responsible for maintaining the filter system — cleaning it, replacing the filter regularly, and storing filtered water appropriately — and not everyone is scrupulous about doing that. A system that isn't maintained properly can add more contaminants to your drinking water than it removes.

The most common type of system is the jug filter, which usually consists of a plastic jug with a replaceable cartridge filter in the upper portion. A jug filter will cost about US$10-$30, and a year's supply of replacement filters will run US$28-$78, according to a 1999 *Consumer Reports* survey. The top portion is filled with tap water which is filtered as it trickles through the cartridge and into the main portion of the jug. Jug filters will usually remove most metals, organic substances, chlorine and water hardness. Some filters can reduce *Cryptosporidium* oocysts — look for "NSF standard 53 for cyst reduction" on the label.

Jug filters must be cared for properly to provide a safe supply of water (see *Tips on the Use of Jug Filters,* page 99). Be aware: most filters remove chlorine residuals from tap water. Although chlorine residuals can give the water an unpleasant taste, they inhibit bacterial growth. Once you have filtered chlorine residuals out of the water, bacteria can grow quickly. For this reason, you should always store filtered water in the fridge. Also, filters must be replaced regularly — check the manufacturer's instructions. Overused filters can actually release heavy metals and bacteria into your water, so don't rely on signalling systems to tell you when the filter needs to be changed. According to a *Consumer Reports* test, these systems were not particularly accurate. Some manufacturers will recycle used cartridges. For example, Brita Canada recharges the ion exchange beads and activated carbon so they can be used for industrial purposes and recycles the plastic casing.

Some filters are impregnated with silver, which acts as a disinfectant. However, it has only limited effectiveness, so you should still take precautions to reduce bacterial growth.

Many people use filter systems or jug filters simply to improve the taste of their tap water. If this is your reason for filtering and you can afford the cost, an activated carbon system should meet your needs. Look for a system that meets ANSI/NSF Standard 42 — Drinking Water Treatment Units — Aesthetic Effects for taste and odor reduction.

If you're thinking of buying a filter to improve the quality of your tap water, find out what's in your water first (see Chapter Ten: *If You're Not Happy*). Are there chlorine residuals? Too many heavy metals? Solvents? Buy a filter system that is designed to remove the particular contaminants you're concerned about. Look for filters that meet ANSI/NSF Standard 53 — Drinking Water Treatment Units — Health Effects, which cover 37 different contaminants including cadmium, mercury, lead, PCBs, atrazine, simazine, 2,4-D, toluene, benzene, and radon. Be aware that manufacturers may make claims that fall outside the standards, and therefore have not been verified by the certifying body.

Tips on the Use of Jug Filters

- *Always follow the manufacturer's instructions.*
- *Don't leave water standing for several days – this encourages bacterial growth.*
- *Keep filtered water in the fridge to discourage bacterial growth.*
- *Clean the jug and reservoir weekly.*
- *Don't keep a filter longer than advised by the manufacturer.*

You might also consider "in-line" or "point of use" filter systems that can be installed under your sink to treat larger volumes of water. These offer more convenience than jug filters — you just turn on your tap to get filtered water instead of constantly refilling jugs. They do cost a little more than jug filters, although the filters do not have to be replaced as frequently. A faucet-mounted filter will cost roughly US$17 to $37, and a year's supply of replacement filters can range from US$27 to US$90. Chapter Twelve provides more details about point-of-use filters systems in the context of private water supplies.

CHAPTER TWELVE

If You're Not on the Mains

NOT EVERYONE IS CONNECTED TO THE MAINS; approximately 23 million Americans and 4 million Canadians depend on private supplies for their drinking water. Private wells are particularly common in Prince Edward Island and Nova Scotia, where they serve approximately half the population.

If you need to install or maintain a private water system, this chapter is for you. It describes how to collect water from various sources, how to store and disinfect it, and how to treat it. Some of this can be done yourself if you have the basic skills, but some of it will require a specialist contractor.

OBTAINING AN INDEPENDENT SUPPLY

If you want an independent supply, the key requirements to keep in mind are adequate quantity, adequate quality, and acceptable cost. There are three potential sources of water to choose from: groundwater, surface water and rainwater.

The first step is to calculate how much water you'll need. A potential water source should provide you with enough reliable water to meet all your needs in the house, garden and barn. Plan on a minimum of 37 gallons (140 liters) per person per day for household use. At least 1.3 gallons (5 liters) will be used for drinking and cooking, so it must meet drinking water standards. Of course, it's relatively easy to reduce the amount of water you use around the house, but at this point you're better off to overestimate your needs than to underestimate them (see *Reducing Water Consumption in the Home*, Chapter Nine). Don't forget about guests! You should also take droughts into account — check your local rainfall patterns to figure out how long you'll need to survive without rainfall to replenish the supply. Based on these calculations, decide which source of water can give you the volume you'll need, day in and day out. Groundwater is going to be the most reliable source of water, if your well is deep enough.

Next, consider the quality of water available. A potential source should also provide you with raw water that is reasonably clean. It is usually easier and cheaper to use good quality water that is more difficult to abstract (see Glossary) than to treat polluted water that is easier to obtain. In terms of purity, groundwater is better than surface water or rainwater. The initial costs of accessing a groundwater source are high, but the water you get will probably require much less treatment than water from other sources.

If you choose groundwater, make sure your well or springbox is located away (and preferably uphill) from any obvious source of pollution, including septic tanks, leachfields, animal pens and feedlots, and areas where pesticides or fertilizers are stored or mixed.

If you choose surface water, water sources from upland areas will probably be better than sources from lowland areas (unless the upland area is peaty or used for agricultural purposes). Don't assume that your water will be high quality just because the stream or lake is clear and sparkling. There may be deer or cattle grazing upstream, a mining operation in the headwaters, or a leaking septic tank nearby. Keep in mind that clean water is vital to your health — it's not worth taking risks to save a few dollars.

Finally, check the legalities of using the source. Depending on where you live, you may need an abstraction license, or local authorities may insist that the well is installed by a certified well driller. Your nearest rural extension service is a great resource for all aspects of developing and maintaining a private water supply.

Groundwater

The biggest problem with using groundwater is finding it in the first place. You'll need to have some test holes drilled to discover how far down the water is, how much of it there is, and how good the quality is. Check the telephone book for a contractor who will do this initial assessment. A dowser may also be helpful in choosing the best places to drill test holes.

WELLS AND BOREHOLES

Most private water supplies in North American depend on wells or boreholes. A well has a diameter greater than 40 inches (1 meter) and is dug by hand or mechanical excavation. The walls are made of impermeable brick or cement, and the water comes up through a layer of sand at the bottom of the shaft (see Figure 12.1, page 104).

A borehole is smaller and deeper than a well, and is drilled by a contractor with special equipment. The borehole is lined with a perforated steel shaft, and water flows in through the holes (see Figure 12.1, page 104).

Wells and boreholes can be classified by depth. Shallow wells or boreholes abstract water from water-bearing rocks above the first impermeable layer (see Figure 12.2, page 105). These are less reliable and can dry up during droughts, and they are more susceptible to pollution. Deep wells or boreholes abstract water from water-bearing rocks below an impermeable layer. They provide a more reliable supply and are less likely to be contaminated.

There are many different types of wells and boreholes, including dug, bored, driven, jetted and drilled. Which type you choose will depend on the resources you have available, the type of soil you'll need to dig through, and how far you'll need to dig to hit a reliable source of water. A contractor can help you make the appropriate choice.

DUG WELLS

If there's water close to the surface, you can dig a well with a backhoe and then line it with cement rings or brick. Or, if the soil is soft enough, you can dig it by hand. Take a concrete ring — say 3 feet (90 centimeters) in diameter — and place it on the ground. Stand inside the ring and start digging. As you remove the earth, the concrete ring will move down. Eventually, when the top of the first ring is level with the ground, you mortar a second ring on top and continue digging. Keep digging and adding rings until you're well into the water table. Then line the bottom of the well with a layer of sand about 20 inches (50 centimeters) deep.

Dug wells can be up to 50 feet (15 meters) deep. They are suitable for clay, silt, sand, gravel, boulders, soft sandstone, and soft fractured limestone; they are not suitable for dense igneous rock. There are a couple of disadvantages to dug wells: they tend to have low yield and can run dry in the summer, and they are difficult to protect from pollution.

Figure 12.1 Wells and Boreholes

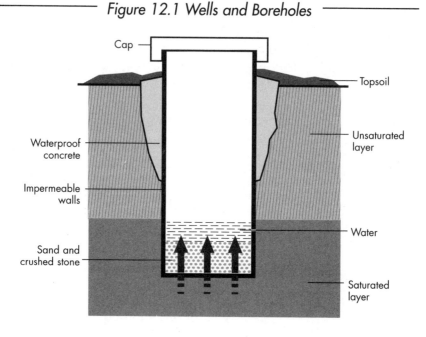

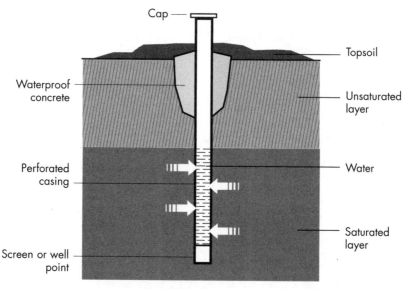

Wells (top) are much wider than boreholes and frequently more shallow. Water enters a well from the bottom, often through a layer of sand and crushed stone. Water enters a borehole through the perforated casing. Note that these diagrams are not to scale.

Figure 12.2 Classification of Wells by Depth

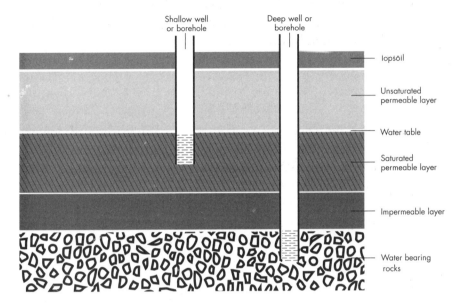

Shallow wells and boreholes are less reliable than deep ones, because they can dry up if the water table drops. Note that this diagram is not to scale.

If the water is further down, you'll probably need to hire a contractor with specialized drilling equipment. This will cost a few thousand dollars, but will produce a deeper well that is less likely to become contaminated and will provide a more reliable supply.

BORED WELLS

These are deeper than dug wells (up to 100 feet or 30 meters), and are lined with tiles, concrete pipe, wrought iron or steel casing. Diameters range from 2 to 30 inches (5 to 75 centimeters). Like dug wells, they are suitable for clay, silt, sand, gravel, boulders smaller than the diameter of the well, soft sandstone, and soft fractured limestone; they are not suitable for dense igneous rock. They are also susceptible to pollution and may not provide a reliable supply of water.

DRIVEN WELLS

This is the least expensive and simplest type of well to construct. A special well point is attached to a steel pipe and driven into the ground. Driven wells can be up to 50 feet (15 meters) deep and 1 to 2 inches (3 to 5 centimeters) in diameter. They are suitable for clay, sand, fine gravel, and sandstone in thin layers; and unsuitable for cemented gravel, boulders, limestone, or dense igneous rock. Like dug and bored wells, they can be prone to pollution and may not provide a reliable supply.

Jetted Wells

Jetting is quick and effective in soft, loose soil. Water is forced down a drill pipe by a pressure pump and out through holes in the drill bit, pushing cuttings to the surface through the space between the casing and the drill pipe. Jetted wells are up to 100 feet (30 meters) deep and 14 to 12 inches (10 to 30 centimeters) in diameter. They are suitable for clay, silt, sand, and small pea gravel; they are not suitable for cemented gravel, boulders, sandstone, limestone, or dense igneous rock.

Drilled Wells

Drilled wells provide the safest and most reliable supply of water and are excavated using machine-operated drilling rigs or equipment. Suitable for clay, silt, gravel, cemented gravel, boulders (in firm bedding), sandstone, limestone, and dense igneous rock, they can be up to 1,000 feet (300 meters) deep and 4 to 24 inches (10 to 60 centimeters) in diameter.

Before you begin to dig, check the regulations. Locate the well so that you minimize the threat of surface drainage getting into the well, and ensure that it's accessible for maintenance and repair. Dig during the driest time of the year when the water table is lowest to make digging easier and reduce the pumping costs.

To prevent surface runoff from contaminating your water supply, make sure there's at least 8 inches (20 centimeters) of casing above the ground, and seal around the casing with waterproof concrete. Keep the well or borehole securely covered, and don't let in any light — light can encourage the growth of algae and micro-organisms in your water supply.

Once it's built, you'll need a pump to get the water from the well. For a borehole, choose a submersible pump that sits inside the borehole below the water level. For a well, there are several choices. You could use a submersible pump, an in-line pump (located on the supply pipe), or a hand bucket. An in-line pump has the advantage that it's more accessible for repairs.

When you call your supplier, you should know the volume of water you'll want to pump each hour, the head (the height difference between your supply and your storage tank) and the bore of the pipe through which you'll be pumping the water. The supplier can then advise you on the best model for your needs.

Finally, you'll need to disinfect your well or borehole. This is not the same thing as the ongoing process of disinfecting your water supply — it is a one-time job to clean the well or borehole itself. The easiest way to do it is with chlorine tablets (calcium hypochlorite) which you can get from your local contractor. Drop them down the well and let the chlorinated water sit for 24 hours. You can also disinfect a well with ordinary household bleach (see *Disinfecting Your Well* below). Once the well is disinfected, pump out all the chlorinated water (you'll know when you're finished when the water stops smelling of chlorine). Now you're ready for business.

If you choose to hire a contractor, all this work will probably be done for you. A contractor will generally take care of everything from drilling, installing the well casing, sealing the well, pumping it, and disinfecting it, up to piping the untreated water to your house. The only things you'll be responsible for are setting up the treatment system, connecting the supply to your household plumbing, and, of course, the ongoing maintenance.

Well maintenance is not a lot of work. You should check the water level periodically to make sure it's not going down, and check your well seal and casing regularly for any signs of cracks, leaks or erosion. The most common sources of well contamination are cracks in the well seal, problems with the well casing, waste from pets or livestock, agricultural chemicals, road salt, and leaking or malfunctioning

Disinfecting Your Well

You can use ordinary household bleach to disinfect your well. Choose a brand that is unscented and contains 5.25 percent sodium hypochlorite. It's best to buy it fresh, because bleach can lose half its strength in just six months. Before you begin, remove or bypass any carbon filters in the system.

Then add the bleach as follows:

- *For a dug well, add one quart for every five feet of water depth (1 liter for every 1.5 meters).*

- *For a drilled well, add 5 ounces for every 25 feet of water depth (142 milliliters for every 7.5 meters).*

- *For a well point, add 3 ounces for every 10 feet of water depth (85 milliliters for every 3 meters).*

septic systems. Most importantly, you should test the water quality at least once a year, or preferably once a season (see *Testing Your Water Supply*, page 122).

Don't use pesticides or fertilizers in the area around the well, and don't install dog runs nearby. Make sure garbage or manure piles don't drain towards the well casing. And if you're on a septic system, make sure it's functioning properly! Don't pour fats, paints, antifreeze or antibacterial products down your drain — they'll kill the bacteria that make your septic system work. For the same reason, you should use biodegradable cleaning products. Finally, never use abandoned wells or dry wells to dispose of garbage or wastewater.

If you need to do any major repairs on your well for any reason, be sure to disinfect it afterwards.

Springs

Groundwater can also be obtained from a spring — a point where groundwater flows out of a crack in the aquifer to the surface. Some springs are reliable all year round, while others can dry up after short periods without rain. Make sure your spring is reliable before you decide to use it as your water supply.

Springs must be protected from pollution. A shelter will ensure that water flows directly from the spring into your pipe without the opportunity for contamination (see Figure 12.3, page 109). After the shelter is built you should disinfect it with chlorine tablets or household bleach. Then the only maintenance required is regularly inspecting the spring box for any cracks, cleaning out the silt once a year, and testing the water quality regularly (see *Testing Your Water Supply*, page 122). It's a good idea to fence the area around your spring box to keep livestock away. As well, keep your cover locked to keep out unwanted visitors.

Surface Water

Surface water comes in a variety of forms, including streams, rivers, ponds and lakes. It's easier to find and use than groundwater, but it is more vulnerable to pollution. Before selecting a potential water source, consider whether the water quality is acceptable. Are there sources of contamination upstream? Farms are one of the biggest sources of contamination, through runoff of fertilizers, pesticides and animal wastes. Industrial effluent and sewage outfalls are other obvious sources of pollution, but also check for abandoned mines. You may need to get the water quality tested (see Chapter Ten: *If You're Not*

If You're Not On The Mains 109

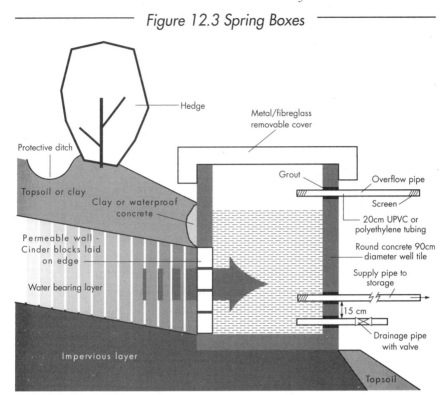

Figure 12.3 Spring Boxes

A spring box will protect your spring from contamination. The box is built into the hillside, with a permeable rear wall that allows the water to flow in. A protective ditch and hedge are located above the box to divert surface runoff. Note that this diagram is not to scale.

Happy). Finally, how reliable is the source? Will it dry up in summer? How much water will be lost through evaporation and seepage?

To obtain water from a surface source, you may need to dam the supply, or you may simply be able to stick a 1-inch (25 millimeter) bore pipe into the source to carry the water to a storage facility below. The intake pipe should be covered by a mesh screen or bars to prevent leaves, twigs and assorted aquatic life from getting into your supply. You can construct a bulb out of wire mesh and stick it on the end of the pipe — this is less likely to get clogged up by a few leaves than a flat screen would be.

Make sure the pipe is fixed firmly in place to prevent it from washing away. The mouth of the pipe should be well above the river bottom so that it's in no danger of being silted up or buried, but well below the water surface to make sure your intake isn't left high and dry if the water level drops. A rocky site is your best bet. If you're drawing water from a lake, keep your intake below the layer of warm water that sits on top of the lake to ensure your water is as clean as possible — warm water encourages the growth of algae.

The next component of your surface water system is a water line running from your intake to your pump house. Your water line is vulnerable to freezing and cracking, so protect it as much as possible. Keep your pump house locked, and if winter temperatures drop below freezing, your pump house should be insulated.

Since surface water tends to have a lot of turbidity, it's an excellent idea to create a settling pond that allows particles to settle out of the water before you treat it. The cleaner your water is when it reaches your treatment system, the more effective that treatment will be.

As far as maintenance goes, you'll need to make sure your intake doesn't get clogged. In the fall, this may mean checking it several times a week when all the leaves come down. You should also inspect your water line regularly for any holes or cracks, and, of course, have your water tested regularly (see *Testing Your Water Supply*, page 122).

Rainwater

Rainwater is readily available and easy to collect from your roof. However, it can accumulate a lot of contaminants, both from pollution in the air and from the surface it falls on. It's not a good idea to rely on rainwater in polluted urban areas, because the rain will pick up a lot of contaminants from the air, particularly hydrocarbons, which are difficult to remove. Also, because it depends directly on rainfall, it is not as reliable as other sources. However, having said all this, our case study demonstrates it can be done effectively (see *Case Study: Rainwater Collection*, page 111). There are roughly 200,000 rainwater cisterns in use in the United States, and California offers a tax credit for rainwater harvesting systems.

To estimate how much water you can collect, you need to know your annual rainfall and the collecting area of your roof. Don't be fooled — the collecting area is not the number of square feet of roofing, but the "footprint" of your roof — the number of square feet it would occupy if it sat directly on the ground. Figure that you can collect 0.5618 gallons for every inch of rain that falls on a square foot of roof collecting area. (To do the metric calculations, count on 0.8 liters for every one millimeter of rain that falls on a square metre of roof collecting area.) So to calculate your average daily yield, use the following formula:

[annual rainfall (inches/year) x collecting area of roof (ft^2) x 0.5618 (gallons/inch/ft^2)] ÷ 365 days/year = average gallons/day.

or:

[annual rainfall (mm/year) x collecting area of roof (m^2) x 0.8 (L/mm/m^2)] ÷ 365 days/year = average L/day.

To prevent contamination, make sure all the surfaces the rain will contact are clean, smooth and non-toxic. Rough surfaces collect dirt and debris that will wash off with the rain. Make sure your eavestroughs and downpipe system are made of materials approved for potable water systems. Cover your eavestrough with screen or mesh to prevent leaves from blocking them up, and use a screening tank before storage.

CASE STUDY: Rainwater Collection

The Toronto Healthy House relies on rainwater collection as its sole source of water. This innovative urban house was the 1992 winner of Canadian Mortgage and Housing's Healthy Housing Design Competition, which looks for innovative approaches to occupant health, energy efficiency, resource efficiency, environmental responsibility and affordability.

The 1,700-square-foot, three-bedroom house has an entirely self-sufficient water supply. Eavestroughs channel rainwater and snow that fall on the roof through screen filters and into a 20,000 liter (5,280 gallon) underground cistern containing limestone to reduce acidity. The cistern is large enough to hold a six-month supply of water. From there, the rainwater is purified with a combination slow sand and activated carbon filter, followed by UV disinfection. Treated water is then stored in cold and hot water tanks that feed the kitchen and bathroom sinks. (see Figure 12.4, page 112).

The Toronto Healthy House uses a tenth of the water of a typical three-person house, thanks to water-efficient fixtures and to greywater recycling. Wastewater is treated using microorganisms, oxygen, UV light and charcoal, and then reused in toilets, showers and the washing machine.

Figure 12.4 Case Study: Rainwater Collection

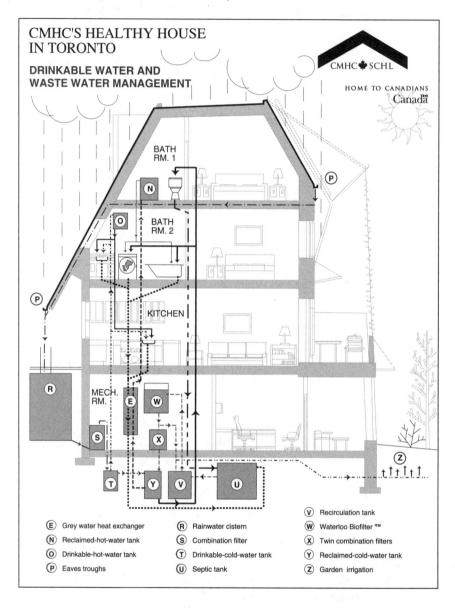

Rainwater can be collected in a cistern underneath the downspout (see Figure 12.5, page 113). Most cisterns are reinforced concrete tanks with a covered access hutch. It's a good idea to keep this access cover locked to prevent accidents, and tightly sealed to prevent contaminants or light from getting in.

Figure 12.5 Rainwater Cistern

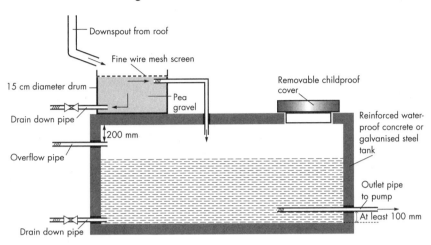

The filtering box directly underneath the downpipe collects the first flush of rainwater, which can contain a lot of contaminants. Once the box is full, water flows across the top of this and into the main tank, where it is stored before treatment. After a rainfall, the filtering box should be drained out. Note that this diagram is not to scale.

The size of your cistern will be determined by your daily water consumption and your local rainfall patterns. You'll need to ensure you've got a big enough supply to get you through any droughts. In the US, the National Weather Service can advise you on long-term rainfall patterns in your area. In Canada, check with Environment Canada's Weather Service.

Disinfect the cistern with chlorine tablets or household bleach after it's been installed, and always divert the first flush of rainwater from the cistern, since it will be the most contaminated. A filtering box will do the trick nicely (see Figure 12.5, above). As far as ongoing maintenance, you'll need to clean out the sludge that accumulates at the bottom of your cistern, check that the covering is intact, and disinfect the cistern on a regular basis. Drain the filtering box after each rainfall. Keep your eavestroughs clear, especially in the fall. And like any water system, test the water quality regularly (see *Testing Your Water Supply*, page 122).

Rainwater is soft. It's good for bathing, laundry and dishwashing, but it is not as healthy to drink as hard water. To make it suitable for drinking, you can add calcium carbonate to your supply to harden it. A lump of chalk rock about the size of your fist will do the trick, or you can buy commercial hardeners.

STORING AND DISTRIBUTING YOUR SUPPLY

Storing water serves two functions: it buffers the differences between supply and demand, and, if the water store is higher than your tap, it can produce pressure in the system.

The size of your storage tank will depend on demand (see *Obtaining an Independent Supply,* page 101). It must be able to meet your peak demand (probably half of your total daily requirement is used during a few peak hours).

The mains pressure at conventional taps is approximately 20 pounds per square inch (psi), with a flow rate of 0.13 gallons per second (0.5 liters per second), but you can probably get by with a little less. Two feet (600 vertical millimeters) equals 1 psi. of pressure; so to create mains pressure you'll need to raise your water up 40 feet (12 meters). But water is heavy, and most houses are not structurally designed to support large tanks of water in the attic. You could use a pump to create pressure directly, but the more efficient method is to have your main storage tank at or below ground level, and use a pump to raise some of it up to a smaller tank in the attic (see Figure 12.6, page 115).

Storage tanks should be made of clean, inert and non-porous material, for example bricks with an impervious render, ferro-cement with render, or plastic containers (plastic barrels used commercially for fruit juice are a good choice). The tank should be covered, but easily accessible.

Piping can be steel, PVC, copper or polythene. The size of bore will depend on how much water you're dealing with. In rough terms, the pipes that deliver water from your source to a storage tank should have a 0.75 to 1-inch (19 to 25-millimeter) bore; the pipes that move water from storage tank to storage tank should have a 0.25 to 0.3-inch (6-8-millimeter) bore, and pipes that deliver stored water to your tap should have a 0.04 to 0.08-inch (1 to 2-millimeter) bore.

To complete your system you'll probably want a float switch to switch the pump on when water levels become too low and off when water levels become too high, or a pressure-operated pump that switches on in response to reduced pressure when you turn on the tap.

Figure 12.6 Sample Household Storage System

[Diagram of a house cross-section showing: Small elevated tank in the attic; Elevated tank provides water pressure; Pump forces water up; Main storage tank in the basement; External supply entering from the right; Pump at the base of the storage tank.]

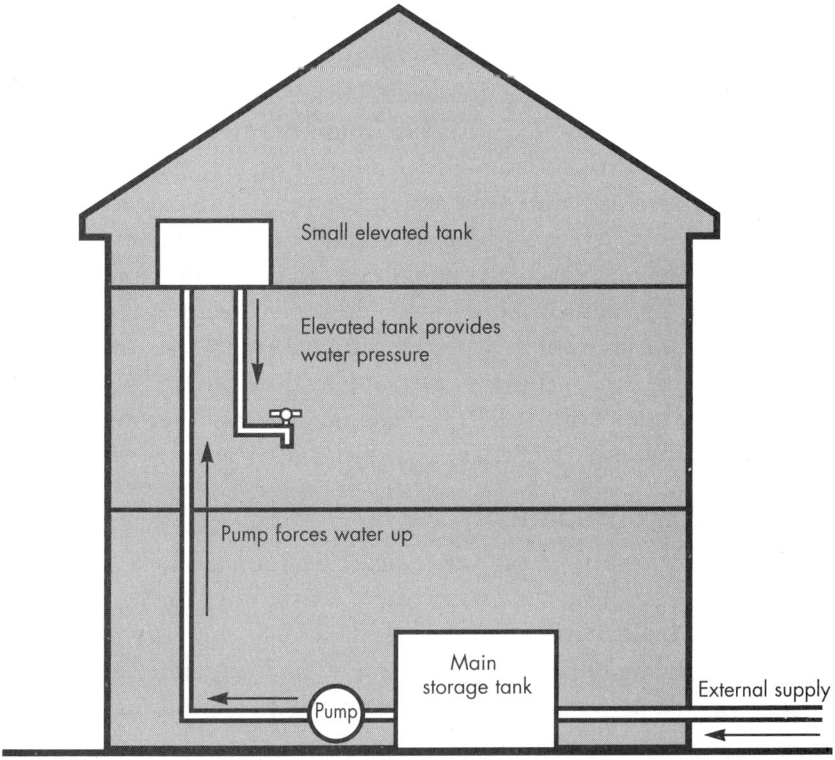

In a typical household storage system, the main storage tank is located in the basement. A portion of the water is pumped up to an elevated tank in the attic to create water pressure at the taps. Note that this diagram is not to scale.

MAKING IT DRINKABLE

Once you've decided on a water source and you've set up your water supply system, it's time for the most important step — making sure the water is safe to drink. The degree of treatment you'll need depends on the quality of the raw water. You should first have the raw water tested to decide what treatment is required. Once the treatment system is operating, test the treated water regularly to make sure the treatment is effective (see *Testing Your Water Supply*, page 122). The quality of surface water can change quite a bit from day to day, so have it tested several times to get a clear picture of how much treatment will be necessary.

Generally speaking, groundwater is high-quality water and does not require as much in the way of treatment. Surface water requires more extensive treatment because it can be polluted by various sources. At minimum, it should be filtered and disinfected, but base your treatment regime on the worst expected raw water quality. Rainwater can also be quite polluted, so disinfection is essential.

It may be worth installing two separate water systems: one for high-quality drinking water, and one for lower-quality water for non-potable purposes such as flushing toilets and washing clothes. This can save you money on treatment, but will require a dual plumbing system.

There are two broad categories of water treatment: point-of-entry treatment and point-of-use treatment. Point-of-entry systems are suitable for large private water supplies (such as for small communities), while point-of-use systems are best for individual household supplies.

Point-of-Entry Treatment

Point-of-entry systems treat water near the source. The system you choose will depend on the quality of your raw water.

You can build yourself a basic slow sand filter that will remove most bacteria, insoluble organic pollutants, and suspended solids (see Figure 12.7, page 117). Choose a container with a rough interior (for example, concrete blocks rendered on the inside) to prevent water from running down the sides without being filtered. Then fill it with layers of particles starting with pea gravel and finishing with very fine sand (sharp sand intended for cement is suitable and can usually be found at a building supply store). It helps to wash the sand first to remove any dust or fine particles that will cause long start-up times. You should end up with a depth of 3 to 4.25 feet (0.9 to .3 meters).

Once it's in use, a layer of algae and other organisms will begin to grow on top of the sand. This is the "schmutzdecke," and it helps to filter the water. Maintenance is very simple: the schmutzdecke and the first few centimeters of sand must be cleared off every few months when it starts to clog things up, and the sand should be topped up every five to six years so it's always 3 feet (90 centimeters) deep. A 10.75 square foot (1 square meter) sand filter will treat enough water for 4 to 10 people, and this is also the smallest effective filter you can build (even if your household consists of just one person).

If you install an in-line flow meter and valve on the outlet pipe, you can monitor the water flow. Ideally the flow should be between

Figure 12.7 A Simple Slow Sand Filter

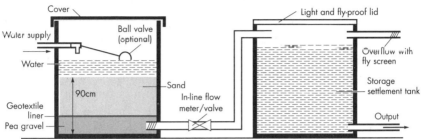

This simple filter works on the principle of slow sand filters in municipal water treatment plants. The water is filtered by the sand itself, and by the layer of micro-organisms that develops on top of the sand. A nylon curtain makes a good geotextile liner. Note that this diagram is not to scale.

2.45 and 9.8 gallons of water per square foot of sand per hour (0.1 to 0.4 cubic meters of water per square meter of sand per hour) to ensure good treatment. When it drops below this flow rate, open the valve a little more. When the valve is fully open and the flow is less than 2.45 gallons per square foot per hour (0.1 cubic meters per square meter per hour), it's time to clear off the schmutzdecke. Make sure there's always water trickling through the filter so the sand doesn't dry out. It's a good idea to use an in-line microfilter after the sand filter to catch sand particles that might wash out in the filtered water and cause wear and tear on your plumbing and fixtures.

If your raw water is relatively clean, the slow sand filter may be all you need, although disinfection is a wise precaution. If your water is very silty or cloudy, you should settle out the suspended particles in a simple settling tank located before the sand filter. If there's lots of bacteria in the raw water, you should definitely disinfect the water after it's been filtered. You can buy UV disinfection units or automatic chlorinators for this purpose.

Point-of-Use Treatment

These off-the-shelf systems are appropriate for generating relatively small volumes of drinking water. They can be used to treat a private water supply, or to improve the quality of mains water. Different models are available: some are installed before the tap, some connect directly to the tap, and some stand separate from the plumbing system.

Many point-of-use systems are connected to a UV disinfection unit to kill any bacteria remaining after filtration. Choose the system that will remove the specific contaminants your incoming water contains (see Figure 12.9, page 120, as well as *Drinking Water Characteristics*, Chapter Two for descriptions of the most common contaminants), but keep in mind that a point-of-use system may not be adequate if the incoming water is too contaminated. Look for a system that meets the relevant ANSI/NSF standards, and make sure it will be able to meet your peak demands.

Particulate Filters

These are used to reduce turbidity or remove specific inorganic particulates such as iron, aluminum or manganese compounds. They may also remove bacteria, depending on pore size. There are several different types available: disc, woven or resin bonded cartridges, and ceramic candles. Bacteria may grow on particulate filters, so there is a danger they could contaminate the treated water. Some filters are impregnated with silver to prevent or inhibit bacterial growth.

Activated Carbon Filters

Activated carbon filters work by adsorbing contaminants onto the surface of carbon particles. They can improve the taste and odor of your water by removing certain industrial wastes, pesticides, decaying organic material, dissolved gases, residual chlorine and by-products of chlorination. They will also remove (to varying degrees) suspended solids, turbidity and organic contaminants. There are two basic types: cartridge filters and activated carbon bed filters.

The cartridge models can fit on a tap or mount under the sink. The activated carbon is contained in replaceable cartridges — granular activated carbon is most common, but powdered activated carbon and block carbon can also be used. A particulate filter at the outlet of the cartridge removes carbon particles from the treated water.

The bed model is designed to treat large volumes of household water (see Figure 12.8, page 119). It is placed after the disinfection unit and the water softener, if these are present. Activated carbon bed filters must be backwashed periodically.

One concern about activated carbon filters is that, by removing chlorine residuals, they may allow bacteria to grow — some studies have shown that water leaving a point-of-use activated carbon filter system may contain more bacteria than untreated water. To prevent

Figure 12.8 Activated Carbon Bed Filter

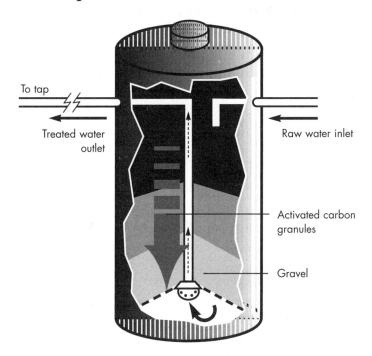

Activated carbon bed filters are often installed underneath the sink. The activated carbon granules remove contaminants as the water filters through. The treated water then goes directly to the tap. Note that this diagram is not to scale.

---

this, some filters are impregnated with silver; others are close-coupled to a UV disinfection unit.

PRE-COAT ACTIVATED CARBON FILTERS

These systems are similar to activated carbon filters, but instead of using granular carbon, they use powdered carbon. They provide more effective filtration. Like other filter systems, they may encourage bacterial growth by removing any chlorine residuals present in the water.

REVERSE OSMOSIS UNITS

Reverse osmosis will remove (to varying degrees) dissolved inorganic contaminants such as sodium, calcium, nitrates and fluoride, and organic contaminants including pesticides and solvents. The functional

Figure 12.9 Comparison of Point-of-Use Systems Effectiveness at Removing Contaminants

	Particulate filters	Activated carbon filters	Reverse osmosis units	Ion exchange units	UV disinfection units
Suspended solids	yes	some	yes	no	no
Hardness	no	no	yes	yes	no
Nitrates	no	no	yes	yes	no
Lead	no	depends on filter type	yes	yes	no
Bacteria, Viruses & cysts	some	some	yes	no	yes
Chlorinated organic compounds	no	yes	yes	some	no
Tastes & odors	no	yes	yes	some	no

basis of a reverse osmosis unit is a semi-permeable membrane, usually made of polyamide, that allows water but not other molecules to pass through (see Figure 12.10, page 121). A high pressure pump pushes water through the membrane at a force of 200 to 420 psi (14 to 29 bar). Salts, minerals, dirt particles, bacteria and viruses are trapped by the membrane. Incoming water constantly cleans the membrane, and contaminants are washed out in the "reject" water.

Although reverse osmosis systems are very good at removing contaminants, they do have several drawbacks. First of all, the quality of the incoming water must be fairly high to prevent the membrane from scaling and fouling, so some kind of pretreatment may be necessary if your raw water isn't very clean. The membrane itself must be chemically cleaned and replaced every few years. Flow rate is very slow, so the treated water must be collected in a storage tank to buffer supply and demand. The treated water is often soft and acidic, making it not as healthy to drink. Finally, for every one volume of clean water produced by the system, three volumes are wasted as "reject" water.

Figure 12.10 Reverse Osmosis Filters

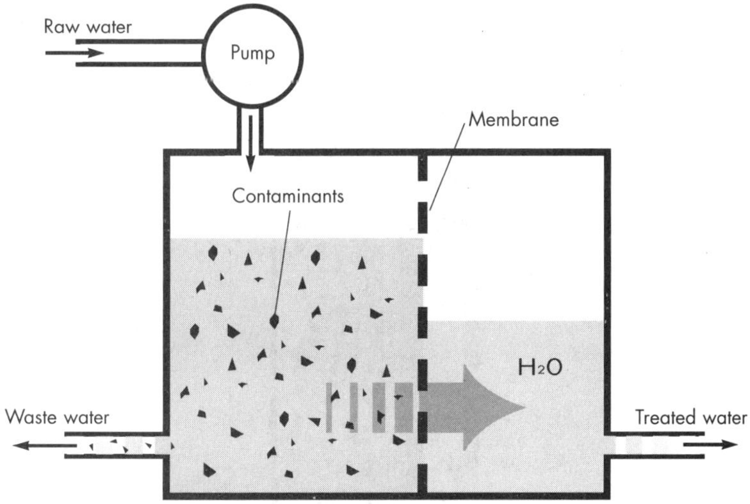

Raw water → Pump
Membrane
Contaminants
H₂O
Waste water ←
Treated water →

The heart of a reverse osmosis unit is a semi-permeable membrane that allows water but not contaminants to pass through. Pressure created by the pump pushes water across the membrane. Note that the diagram is not to scale.

ION EXCHANGE UNITS

Ion exchange units do not purify water per se, but they can soften hard water or remove nitrates. These systems use synthetic resins — small beads 0.2 to 2.0 millimeters in diameter. They can be cationic or anionic. Cationic resins soften water by exchanging positively charged calcium and magnesium ions with sodium or potassium (see Figure 12.11, page 122), while anionic resins exchange negatively charged nitrates with chloride ions. When the resins are exhausted, they can be regenerated with sodium chloride solution. Softened water is good for washing, but it's not as healthy to drink as hard water.

ULTRAVIOLET DISINFECTION UNITS

Known as UV units for short, ultraviolet systems will disinfect your water supply by inactivating the micro-organisms that could cause disease. However, they won't remove anything else that might be in the water, such as suspended particles, metals or chlorinated organic compounds, and they won't solve taste or odor problems. They can be a good choice as a final stage of treatment after the water has passed through a filter to remove particulates.

Figure 12.11 Water Softeners

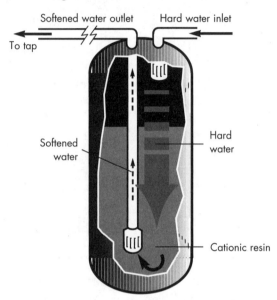

Water softeners work by exchanging calcium and magnesium ions that cause hardness with sodium ions. The exchange occurs in the cationic resin, which must be recharged periodically with saline solution. Note that this diagram is not to scale.

UV systems consist of an array of UV light bulbs (which look similar to ordinary fluorescent bulbs) housed in transparent quartz sleeves. Water is irradiated by light at the ultraviolet frequency as it passes between the bulb and the sleeve.

It's important that the water coming in is relatively clean — the dirtier the water, the less effective UV disinfection will be. The quartz sleeves should be cleaned periodically, and the bulbs will need replacing two to three times a year.

TESTING YOUR WATER SUPPLY

Maintaining your private water supply should not be a lot of work, but you do need to test your water quality regularly. If you're using a well, you should test it at least once a year, and preferably once a season. The same is true for springs. It's also a good idea to test a spring after heavy rains when there is the highest risk of contamination from surface water. Surface water sources should be tested frequently.

In the United States, call your local or state public health department for a list of certified laboratories in your area. Note that in some jurisdictions, government subsidies for private well testing may be available, or testing may even be free. The lab will send you a sample bottle and a list of instructions (see *How to Collect a Water Quality Sample* for an example).

How to Collect a Water Quality Sample

1. *Obtain a water sample bottle from your local health unit or laboratory.*
2. *Do not touch the bottle lip and do not rinse out the bottle.*
3. *Remove aerators and other attachments from your tap. Let the cold water run for two to three minutes before sampling.*
5. *Fill the bottle to the "fill line" directly from the tap without changing the flow of water.*
6. *Replace cap tightly.*
7. *Keep samples refrigerated after collection.*
8. *Complete the form that came with the bottle.*
9. *Return the sample and form to the health unit or laboratory within 24 hours of collection.*

(Source: Green Communities Association, 2002.)

Conclusion

CURRENTLY MOST TAP WATER MEETS North American standards for safe drinking water, so there is no need to panic about the quality of your water. In most cases, the health risks of drinking water are quite small, especially compared to smoking, driving a car, or even breathing urban air. In fact, our drinking water systems are a tremendous accomplishment, both in terms of engineering and public health.

Does that mean they are perfect? Absolutely not. It's worth being concerned about the quality of your tap water. Good water is vital to human survival, and it's not something to be taken for granted.

Start by finding out what's in your tap water. If you live in the United States, read the Consumer Confidence Report that your water supplier should send you each year. In Canada, getting information may not be quite so simple. If your water supplier can't help, try your provincial Ministry of Health or Environment.

So what do you really need to be concerned about? If you're living in a small town that doesn't have the financial resources or technical expertise of larger cities, there's a greater chance of problems with your water quality.

Bacteria, viruses and pathogens are the biggest dangers because they can be fatal. If your water supplier is doing its job right, you don't need to worry. However, if your immune system is compromised for any reason — because of HIV or chemotherapy, for example — there may be enough micro-organisms in your tap water to make you sick. Ask your doctor what steps you should be taking to protect your health, such as boiling your water or using a filter system.

Lead is the other big worry. It's particularly dangerous for young children because it interferes with normal brain development. Check your plumbing or consider having your water tested. If you discover that there's lead in your water, there are home filters that will get it out — look for a system that meets ANSI/NSF standards for lead removal — or consider replacing any lead pipes within your house.

There are many other unhealthy substances that can find their way into drinking water, including pesticides, nitrates, arsenic, radon and disinfection byproducts, to name a few. The proper filter system can get rid of many of these, so it pays to know what you're drinking. That's why your Consumer Confidence Report is important reading.

More and more people are turning to filtered water or bottled water because it is "healthier" or "tastes better." That's too bad. In many cases, their tap water is fine and there's no need for expensive alternatives. In some cases, there are problems with the tap water, and filter systems and bottled water may be good short-term solutions. Keep in mind, however, that these are only short-term solutions for people who can afford it. North America has poured enormous amounts of effort and money into creating public water systems, and we should continue to focus on making them as good as possible for everyone, rather than opting out by buying home filter systems or bottled water if we have the money and ignoring the tap water.

It's also worth remembering that if you choose to filter your water, you must take care of your filter system properly. If you don't, you run the risk of actually increasing the amount of contaminants in your water. Likewise, if you buy bottled water, you should know what's in it. There's no guarantee it's going to be any more pure than what comes out of your tap, so ask the manufacturer for a complete list of what's in there.

Having said all that, it's true that public systems will never be entirely risk free. One of the problems that governments deal with is deciding how much money to spend reducing those risks, and how much to spend dealing with other public health risks like infectious diseases and second-hand cigarette smoke. Is it worth spending billions of dollars to test for a relatively rare contaminant that may or may not have any effect on human health, when government could get more bang for its public health buck by investing in prenatal nutrition programs, for example?

In the United States, the EPA's Safe Drinking Water Act is quite strict, but even so, there are dozens of common drinking water contaminants that aren't covered. One of the biggest criticisms of the EPA's drinking water program is that it's not moving fast enough to establish regulations for high-priority contaminants.

In Canada, the picture isn't so good. Drinking water quality is a provincial issue, and standards vary considerably across the country. Alberta, Quebec and Ontario have strong legislation, but Prince Edward Island, for example, currently has no legally enforceable standards for any drinking water contaminants. Many provinces have taken a fresh look at their drinking water regulations, thanks to recent waterborne disease outbreaks in Walkerton and North Battleford, but there's still lots of room for improvement.

Finally, just having strong regulations is no guarantee your water is safe — in the past year, eight percent of American water suppliers didn't meet health-based standards. In Newfoundland, 66,500 people were forced to boil their water in 2001, and British Columbia has the highest rate of intestinal illness in the country, thanks in part to poor water systems.

For all these reasons, we should be pushing for strict, legally enforceable regulations — particularly in many Canadian provinces that currently rely on guidelines alone — and regular monitoring to make sure suppliers are meeting those regulations. We should also be willing to pay the price for good drinking water. Stricter regulations mean more expensive water, but it's a price worth paying and still far cheaper than buying bottled water.

We should also be concerned about water in a wider sense. Water is essential to us and to all other living creatures. Water is a limited resource, and it is an endlessly recycled resource. It is dangerous and short-sighted to pour wastewater into our lakes and rivers and to use ridiculous amounts of high-quality water to flush our toilets or hose down our driveways. The sensible approach is to prevent pollution from getting into our water supplies in the first place, and to treat our tap water as a valuable resource to be used frugally.

In the final analysis then, is our water safe to drink? Although the answer is yes, it's a qualified yes. For most of us, our drinking water is very good. Every year, however, hundreds of thousands of North Americans get sick and dozens die because of contaminated drinking water. Good water isn't something we can take for granted, and it's something we all need to take responsibility for: governments, industry, water suppliers, and individual citizens. Think about it next time you turn on the tap.

Glossary

Abstraction: The removal of water from surface or groundwater sources for domestic, commercial or industrial use.

Acidic: Having a pH value less than 7.

Adsorption: The process by which one substance is transferred onto the surface of another substance by physical and/or chemical forces.

Aeration: The exposure of water to air, so that air becomes dissolved in the water.

Alkaline: Having a pH value greater than 7.

Aquifer: An underground formation of porous, water-bearing rock, sand or gravel.

Backflow: A situation where the flow of treated water reverses and untreated water is sucked into the distribution system.

Basic: Capable of combining with an acid in water to form a salt and hydroxide ions; very basic water is not palatable.

Biofilm: A layer of bacteria that can form on the interior of water pipes.

Carcinogenic: Having the ability to cause cancer.

Catchment: Any surface area that collects runoff during a rainfall; the area from which individual rivers, lakes or reservoirs collect water.

Cholera: A potentially fatal disease of the small intestine characterized by severe vomiting and diarrhea and caused by a strain of bacterium.

Coagulation: A process in which a coagulant (typically aluminum sulfate) is added to water, causing suspended material to group together into aggregate "flocs" which are subsequently settled out by sedimentation and/or flotation. Used to remove turbidity and color.

Coliform: A group of bacteria whose presence in drinking water may indicate contamination by disease-causing micro-organisms.

Contaminant: Any undesirable physical, chemical or microbiological substance in water.

Cross-connection: Any place in the distribution system where treated water can come into contact with wastewater, sewage or untreated water.

Cryptosporidiosis: A potentially fatal disease caused by *Cryptosporidium* (see *Cryptosporidium*), characterized by diarrhea, headache, cramps, nausea and vomiting.

Cryptosporidium: A type of gastroenteritis-causing parasite, which survives in water supplies in the form of oocysts.

Disinfection: The removal, destruction or inactivation of pathogens in water.

Effluent: The liquid waste from industrial, agricultural or sewage plant outlets.

E. coli: A fecal coliform that indicates water has been contaminated with sewage or manure and may therefore contain dangerous pathogens. There are several hundred strains of *E. coli* and most are harmless to humans, but *E. coli* O157:H7 is potentially fatal.

EPA: The United States Environmental Protection Agency.

Epidemiological data: Data obtained from studying how frequently a particular disease or condition occurs and how it is distributed in the population.

Fecal: To do with waste matter discharged from the bowels.

Fecal coliform bacteria: A group of bacteria found in vast numbers in human and animal feces. Since their presence in water indicates potential fecal contamination, they are considered an "indicator species."

Filtration: The process of passing water through porous material to remove suspended particles.

Giardia: A protozoan found in the feces of infected mammals in the form of resilient cysts and responsible for causing "beaver fever." Symptoms develop one to four weeks after ingestion and include explosive, watery, foul-smelling diarrhea, gas in the stomach, nausea and loss of appetite.

Groundwater: Water held in water-bearing rocks beneath the surface of the earth.

Hardness: A property of water caused by dissolved calcium and magnesium salts. Hard water causes scale formation in pipes and kettles. Soft water, which is low in calcium and magnesium, is often acidic and corrosive.

Indicator species: As it is difficult and expensive to screen water samples for all possible pathogens, one species (usually fecal coliform bacteria) are used as indicators of fecal contamination and therefore the possible presence of pathogens.

IMAC: Interim Maximum Acceptable Concentration. The interim guideline Health Canada has set for a particular contaminant.

Ion: An atom or molecule that has lost one or more electrons to become a positively charged cation, or has gained one or more electrons to become a negatively charged anion.

Leaching: The process by which substances are removed from a solid mass by the action of a percolating liquid, and are carried away in solution or suspension.

MAC: Maximum Acceptable Concentration. The guideline Health Canada has set for a particular contaminant.

MCL: Maximum Contaminant Level. The maximum level of a contaminant that the US Environmental Protection Agency will allow in public drinking water supplies.

MCLG: Maximum Contaminant Level Goal. The level of a contaminant in drinking water at which there is no known or anticipated health threat from that contaminant to a person who consumes the water, according to the EPA.

NOAEL: No Observed Adverse Effect Level. The level at which a contaminant does not have any harmful effects on human health, according to Health Canada.

Oocyst: The dormant form of a protozoan parasite such as *Cryptosporidium*.

Pathogen: An organism that causes disease.

Peak demand: The maximum demand (e.g. for water) during a particular period of time (for example, daily or annually).

pH: A logarithmic measure of the acidity or alkalinity of water. A pH of 7 is neutral, a pH of less than 7 is acidic, and a pH of more than 7 is alkaline.

Potable: Safe to drink; free from disease-causing organisms and toxic substances.

Raw water: Another term for untreated water.

Runoff: Water derived from snow or rain that flows across the surface of land into streams, rivers and lakes.

Sedimentation: A process for removing suspended solids by passing water slowly through a tank, allowing sediments to settle out through the force of gravity.

Surface water: Bodies of water on the surface of the Earth, including streams, rivers, ponds, lakes and oceans.

Suspended solids: Particles of grit, sand or organic material floating (suspended) in a body of water.

TDI: Tolerable Daily Intake. The maximum amount of a contaminant that can safely be consumed on a daily basis, according to Health Canada.

THM: Trihalomethane. A byproduct of chlorine disinfection created when chlorine combines with organic material present in the water.

Toxicological data: Data based on (usually animal) studies of how poisonous a particular substance is.

Treatment technique: A set of procedures defined by the EPA that public water suppliers must follow to ensure that a contaminant is controlled in their drinking water supplies.

Turbidity: A measure of the number of particles present in water.

Typhoid: A bacterial disease characterized by fever, intestinal irritation and a rash on the chest and abdomen.

UV irradiation: A form of disinfection using ultraviolet light, which has a frequency just beyond the violet end of the visible spectrum and can inactivate pathogens in contaminated water.

Water soluble: Having the ability to dissolve in water.

Water table: The level underground beneath which the rocks are saturated with groundwater.

Resources

Government Resources, United States

Environmental Protection Agency
Office of Groundwater and Drinking Water (4601)
Ariel Rios Building
1200 Pennsylvania Avenue, NW
Washington, DC 20460-0003
Safe Drinking Water Hotline:
1 800 426-4791
E-mail Hotline: hotline-sdwa@epa.gov
Website: www.epa.gov/safewater

EPA Region 1: Connecticut, Maine, Massachusetts, New Hampshire, Rhode Island, Vermont
EPA New England, Region 1
1 Congress Street
Suite 1100
Boston, MA 02114-2023
Telephone:
(617) 918-1571 or (800) 372-7341
Website:
www.epa.gov/region1/eco/drinkwater/index.html

EPA Region 2: New Jersey, New York, Puerto Rico, US Virgin Islands
290 Broadway
New York, NY 10007-1866
Telephone: (212) 637-5000
Website:
www.epa.gov/region02/water/wpb/region2.htm

EPA Region 3: Delaware, District of Columbia, Maryland, Pennsylvania, Virginia, West Virginia
US EPA Region 3
Water Protection Division (3WP00)
1650 Arch Street
Philadelphia, PA 19103-2029
Telephone: (215) 814-2300
Website: www.epa.gov/reg3wapd/

EPA Region 4: Alabama, Florida, Georgia, Kentucky, Mississippi, North Carolina, South Carolina, Tennessee
US EPA Region 4
Drinking Water Section
Sam Nunn Atlanta Federal Center
61 Forsyth Street, SW
Atlanta, GA 30303-3104
Telephone: (404) 562-9900
Website: www.epa.gov/region4/water

EPA Region 5: Illinois, Indiana, Michigan, Minnesota, Ohio, Wisconsin
USEPA Region 5
Water Division (W-15J)
77 W. Jackson Boulevard
Chicago, IL 60604-3590
Telephone: (312) 353-2147
Website: www.epa.gov/region5/water

EPA Region 6: Arkansas, Louisiana, New Mexico, Oklahoma, Texas
USEPA Region VI
Water Quality Protection Division
1445 Ross Avenue, Suite 1200
Dallas, TX 75202
Telephone: (214) 665-6444
Website:
www.epa.gov/earth1r6/6wq/6wq.htm

EPA Region 7: Iowa, Kansas, Missouri, Nebraska
US EPA Region 7
Office of External Programs
901 N. 5th Street
Kansas City, KS 66101
Telephone:
(913) 551-7667 or (800) 223-0425
Website: www.epa.gov/region7/water

EPA Region 8: Colorado, Montana, North Dakota, South Dakota, Utah,

Wyoming
US EPA Region 8
999 18th Street, Suite 300
Denver, CO 80202-2466
Telephone: (303) 312-6312 or
(800) 227-8917
Website:
www.epa.gov/region08/water/dwhome/dwhome.html

EPA Region 9: Arizona, California, Hawaii, Nevada, the Pacific Islands
US EPA Region 9
Drinking Water Section (W-6-1)
75 Hawthorne Street
San Francisco, CA 94105
Telephone: (415) 947-8000 or
(866) EPA-WEST
Website: www.epa.gov/region09/water

EPA Region 10: Alaska, Idaho, Oregon, Washington
US EPA Region 10
Drinking Water Unit, Office of Water
1200 Sixth Avenue, OW-136
Seattle, WA 98101
Telephone: (206) 553-1200 or
(800) 424-4EPA
Website: http://yosemite.epa.gov/R10/WATER.NSF

National Lead Information Center
801 Roeder Road, Suite 600
Silver Spring, MD 20910
Telephone: (800) 424-LEAD
Website: www.epa.gov/lead/nlic.htm

Government Resources, Canada

Health Canada
Environmental Health Directorate, Health Protection Branch
0801B3
Ottawa, ON K1A 0L2
Website: www.hc-sc.gc.ca/ehp/ehd/bch/water_quality.htm

Alberta Environment Information Centre
Main Floor, 9920-108 Street
Edmonton, AB T5K 2G8
Telephone: (780) 944-0313
Facsimile: (780) 427-4407
Website:
www3.gov.ab.ca/env/water/index.cfm
E-mail: env.infocent@gov.ab.ca

British Columbia Ministry of Health Planning
1515 Blanshard Street
Victoria, BC V8W 3C8
Telephone: (250) 952-1469
Website:
www.healthplanning.gov.bc.ca/protect/water.html

Manitoba Health
Office of the Chief Medical Officer of Health
4th Floor — 300 Carlton Street
Winnipeg, MB R3B 3M9
Telephone: (204) 788-6666
Facsimile: (204) 948-2204
Drinking Water Safety Information Line:
(877) 859-8929
Website: www.gov.mb.ca/health/publichealth/cmoh/water.html
E-mail: mgi@gov.mb.ca

New Brunswick Department of the Environment and Local Government
PO Box 6000
Fredericton, NB E3B 5H1
Telephone: (506) 453-2690
Facsimile: (506) 457-4991
Website: www.gnb.ca/0009/0003-e.asp

Newfoundland and Labrador Department of the Environment
Water Resources Division
4th Floor, West Block
Confederation Building
PO Box 8700

St. John's, NL A1B 4J6
Telephone: (709) 729-2563
Facsimile: (709) 729-0320
Website: www.gov.nf.ca/env/Env/
waterres/Policies/PolicyList.asp

**Northwest Territories Department
of Public Works and Services**
Stuart Hodgson Building
Box 1320
Yellowknife, NT X1A 2L9
Telephone: (867) 873-7817
Facsimile: (867) 873-0104
Water quality database:
http://aurora.gov.nt.ca/waterq/waterq_
main_menu.asp

Nova Scotia Environment and Labour
5151 Terminal Road
PO Box 697
Halifax, NS B3J 2T8
Telephone: (902) 424-5300 or
(877) 9ENVIRO
Facsimile: (902) 424-0503
Website: www.gov.ns.ca/enla/water

**Nunavut Department of Health
and Social Services**
Telephone: (867) 975-5700
Facsimile: (867) 975-5705
Website: www.gov.nu.ca/Nunavut/English
/departments/HSS

Ontario Ministry of the Environment
Public Information Centre
135 St. Clair Avenue West
1st Floor
Toronto, ON M4V 1P5
Telephone: (416) 325-4000 or
(800) 565-4923
Website: www.ene.gov.on.ca/water.htm
E-mail: picemail@ene.gov.on.ca

**Prince Edward Island Ministry
for the Environment**
Water Resources
11 Kent Street, 4th Floor
PO Box 2000
Charlottetown, PEI C1A 7N8

Telephone: (902) 368-5028 or
(866) 368-5044
Facsimile: (902) 368-5830
Website: www.gov.pe.ca/fae/wr-info/
index.php3

Québec Ministère de l'Environnement
Edifice Marie-Guyart, rez-de-chaussée
675, boulevard René-Lévesque Est
Québec, PQ G1R 5V7
Telephone: (418) 521-3830 or
(800) 561-1616
Facsimile: (418) 646-5974
Website: www.menv.gouv.qc.ca/index-
en.htm
E-mail: info@menv.gouv.qc.ca

Saskatchewan Environment
3211 Albert Street
Regina, SK S4S 5W6
Telephone: (306) 787-2700 or
(800) 567-4224
Water Inquiry Line: (866) 727-5420
Website: www.se.gov.sk.ca/environment/
protection/water/drinking.asp
E-mail: inquiry@serm.gov.sk.ca

**Yukon Department of Health
and Social Services**
Environmental Health Branch
Box 2703
2 Hospital Road
Whitehorse, YK Y1A 2C6
Telephone: (867) 667-8391
or (800) 661-0408 local 8391
Facsimile: (867) 667-8322
Website:
www.hss.gov.yk.ca/prog/eh/index.html
E-mail: environmental.health@gov.yk.ca

Industry Associations

American Water Works Association
6666 W. Quincy Avenue
Denver, CO 80235
Telephone: (303) 794-7711
Facsimile: (303) 794-3951
Website: www.awwa.org

Canadian Bottled Water Association
155 East Beaver Creek Road
Unit 24, Suite 328
Richmond Hill, ON L4B 2N1
Telephone: (905) 886-6928
Facsimile: (905) 886-9531
Website: www.cbwa-bottledwater.org
E-mail: info@cbwa-bottledwater.org

Canadian Water and Wastewater Association
5330 Canotek Road
2nd Floor, Unit 20
Ottawa, ON K1J 9C3
Telephone: (613) 747-0524
Facsimile: (613) 747-0523
Website: www.cwwa.ca

International Bottled Water Association
1700 Diagonal Road
Suite 650
Alexandria, VA 22314
Telephone: (703) 683-5213 or (800) WATER-11
Facsimile: (703) 683-4074
Website: www.bottledwater.org
E-mail: ibwainfo@bottledwater.org

NSF International
PO Box 130140
789 N. Dixboro Road
Ann Arbor, MI 48113-0140
Telephone: (734) 769-8010 or (800) NSF-MARK
Facsimile: (734) 769-0109
Website: www.nsf.org/water.html

Nonprofit Organizations

Campaign for Safe and Affordable Drinking Water
4455 Connecticut Avenue, NW
Suite A300
Washington, DC 20008-2328
Telephone: (202) 895-0420 x135
Website: www.safe-drinking-water.org
E-mail: csadw@cleanwater.org

Environmental Working Group
1436 U Street, NW
Suite 100
Washington, DC 20009
Telephone: (202) 667-6982
Facsimile: (202) 232-2592
Website: www.ewg.org
E-mail: info@ewg.org

Green Communities Association
Box 928
Peterborough, ON K9J 7A5
Telephone: (705) 745-7479
Facsimile: (705) 745-7294
Website: www.gca.ca/water.htm

Natural Resources Defense Council
40 West 20th Street
New York, NY 10011
Telephone: (212) 727-2700
Facsimile: (212) 727-1773
Website: www.nrdc.org/water/default.asp
E-mail: nrdcinfo@ndrc.org

Sierra Legal Defence Fund
214-131 Water Street
Vancouver, BC V6B 4M3
Telephone: (604) 685-5618 or (800) 926-7744
Facsimile: (604) 685-7813
Website: www.sierralegal.org/water.html
E-mail: sldf@sierralegal.org

Wellowner.org
(sponsored by the National Ground Water Association)
Website: www.wellowner.org

Consumer and Do-it-Yourself Resources

Consumer Reports
101 Truman Avenue
Yonkers, NY 10703
Website: www.consumerreports.org
A monthly magazine published by Consumers Union, an independent, non-profit testing and information organization.

Contains useful reviews of many consumer products, including low-flow toilets, bottled water and water filter systems.

Oasis Designs
Website: www.oasisdesign.net
A good source of information on greywater recycling, rainwater collection and composting toilets.

Waterwiser
Website: www.waterwiser.org
A website created by the American Water Works Association that provides a wealth of links to water efficiency information.

Publications

Christensen, Randy, *Waterproof: Canada's Drinking Water Report Card*, Sierra Legal Defence Fund, 2001.

Green Communities Association, *How Well Is Your Well: Homeowner's Guide to Safe Wells and Septic Systems*, Green Communities Association, 2002.

Office of Water, *Water On Tap: A Consumer's Guide to the Nation's Drinking Water*, US EPA, 1997.

Olson, Erik, *What's On Tap? Grading Drinking Water in U.S. Cities*, Natural Resources Defense Council, 2002.

Symons, James M., *Plain Talk About Drinking Water: Questions and Answers About the Water You Drink*, American Water Works Association, 1997.

Appendix A

US and Canadian Standards for Drinking Water

Note that the US standards are enforceable regulations. The Canadian standards, on the other hand, are unenforceable guidelines that each province may or may not choose to adopt.

IMAC: Interim Maximum Acceptable Level

MAC: Maximum Acceptable Level

MCL: Maximum Contaminant Level

MCLG: Maximum Contaminant Level Goal

MRDL: Maximum Residual Disinfectant Level

MRDLG: Maximum Residual Disinfectant Level Goal

TT: Treatment Technique

	United States		Canada
Contaminant	MCLG (mg/L)	MCL or TT (mg/L)	MAC (mg/L)
MICRO-ORGANISMS			
Cryptosporidium	zero	TT 99% removal	
E. coli			none
Giardia lamblia	zero	TT 99.9% removal	
Heterotrophic plate count	n/a	TT no more than 500 bacterial colonies per milliliter	
Legionella	zero	TT	
Total Coliforms	zero	5.0%	zero[1]
Turbidity	n/a	TT <5 NTU	1 NTU
Viruses (enteric)	zero	TT 99.99% removal	

APPENDIX A: *US and Canadian Standards for Drinking Water* 137

Contaminant	United States MCLG (mg/L)	United States MCL or TT (mg/L)	Canada MAC (mg/L)
DISINFECTION BYPRODUCTS			
Bromate	zero	0.010	0.01 (IMAC)
Bromodichloromethane	zero	0.08	
Bromoform	zero	0.08	
Chlorite	0.8	1.0	
Dibromochloromethane	0.06	0.08	
Dichloroacetic acid	zero	0.060	
Haloacetic acids	n/a	0.060	
Total Trihalomethanes	none	0.10	0.1 (IMAC)
Trichloroacetic acid	zero	0.060	
DISINFECTANTS			
Chloramines (as Cl_2)	MRDLG=4	MRDL=4.0	3.0
Chlorine (as Cl_2)	MRDLG=4	MRDL=4.0	
Chlorine dioxide (as ClO_2)	MRDLG=0.8	MRDL=0.8	
INORGANIC CHEMICALS			
Aluminum			0.1 (when added as a coagulant to water treatment)
Antimony	0.006	0.006	0.006 (IMAC)
Arsenic	0	0.010 (as of 1/23/06)	0.025 (IMAC)
Asbestos (fibers >10 micrometers)	7 million fibers per liter	7 MFL	
Barium	2	2	1.0

	United States		Canada
Contaminant	MCLG (mg/L)	MCL or TT (mg/L)	MAC (mg/L)
Beryllium	0.004	0.004	
Boron			5 (IMAC)
Cadmium	0.005	0.005	0.005
Chromium (total)	0.1	0.1	0.05
Copper	1.3	TT Action level=1.3	
Cyanide (as free cyanide)	0.2	0.2	0.2
Fluoride	4.0	4.0	1.5
Lead	zero	TT Action level=0.015	0.010
Mercury (inorganic)	0.002	0.002	0.001
Nitrate (measured as nitrogen)	10	10	45
Nitrite	1	1	
Selenium	0.05	0.05	0.01
Thallium	0.0005	0.002	
ORGANIC CHEMICALS			
Acrylamide	zero	TT	
Alachlor	zero	0.002	
Aldicarb			0.009
Aldrin + dieldrin			0.0007
Atrazine	0.003	0.003	0.005 (IMAC) (includes metabolites)
Azinphos-methyl			0.02
Bendiocarb			0.04
Benzene	zero	0.005	0.005
Benzo(a)pyrene (PAHs)	zero	0.0002	0.00001

APPENDIX A: *US and Canadian Standards for Drinking Water* 139

Contaminant	United States		Canada
	MCLG (mg/L)	MCL or TT (mg/L)	MAC (mg/L)
Bromoxynil			0.005 (IMAC)
Carbaryl			0.09
Carbofuran	0.04	0.04	0.09
Carbon tetrachloride	zero	0.005	0.005
Chlordane	zero	0.002	
Chlorobenzene	0.1	0.1	
Chlorpyrifos			0.09
Cyanazine			0.01 (IMAC)
Cyanobacterial toxins (as microcystin-LR)			0.0015
2,4-D	0.07	0.07	0.1 (IMAC)
Dalapon	0.2	0.2	
Diazinon			0.02
1,2-Dibromo-3-choropropane (DBCP)	zero	0.0002	
Dicamba			0.12
1,2-Dichlorobenzene			0.20
1,4-Dichlorobenzene			0.005
o-Dichlorobenzene	0.6	0.6	
p-Dichlorobenzene	0.075	0.075	
1,2-Dichloroethane	zero	0.005	0.005 (IMAC)
1,1-Dichloroethylene	0.007	0.007	0.014
cis-1,2-Dichloroethylene	0.07	0.07	
trans 1,2-Dichloroethylene	0.1	0.1	
Dichloromethane	zero	0.005	0.05
2,4-Dichlorophenol			0.9
1,2-Dichloropropane	zero	0.005	

	United States		Canada
Contaminant	MCLG (mg/L)	MCL or TT (mg/L)	MAC (mg/L)
Di(2-ethylhexyl) adipate	0.4	0.4	
Di(2-ethylhexyl) phthalate	zero	0.006	
Dichlofop-methyl			0.009
Dimethoate			0.02 (IMAC)
Dinoseb	0.007	0.007	0.01
Dioxin (2,3,7,8-TCDD)	zero	0.00000003	
Diquat	0.02	0.02	0.07
Diuron			0.15
Endothall	0.1	0.1	
Endrin	0.002	0.002	
Epichlorohydrin	zero	TT	
Ethylbenzene	0.7	0.7	
Ethylene dibromide	zero	0.00005	
Glyphosate	0.7	0.7	0.28
Heptachlor	zero	0.0004	
Heptachlor epoxide	zero	0.0002	
Hexachlorobenzene	zero	0.001	
Hexachlorocyclopentadiene	0.05	0.05	
Lindane	0.0002	0.0002	
Malathion			0.19
Methoxychlor	0.04	0.04	0.9
Metolachlor			0.05 (IMAC)
Metribuzin			0.08
Monochlorobenzene			0.08
Nitriloacetic acid (NTA)			0.4
Oxamyl (Vydate)	0.2	0.2	
Paraquat (as dichloride)			0.01 (IMAC)
Parathion			0.05

APPENDIX A: *US and Canadian Standards for Drinking Water* 141

Contaminant	United States MCLG (mg/L)	United States MCL or TT (mg/L)	Canada MAC (mg/L)
Polychlorinated biphenyls (PCBs)	zero	0.0005	
Pentachlorophenol	zero	0.001	0.06
Phorate			0.002
Picloram	0.5	0.5	0.19 (IMAC)
Simazine	0.004	0.004	0.01 (IMAC)
Styrene	0.1	0.1	
Terbufos			0.001 (IMAC)
Tetrachloroethylene	zero	0.005	0.03
2,3,4,6-Tetrachlorophenol			0.1
Toluene	1	1	
Toxaphene	zero	0.003	
2,4,5-TP (Silvex)	0.05	0.05	
1,2,4-Trichlorobenzene	0.07	0.07	
1,1,1-Trichloroethane	0.20	0.2	
1,1,2-Trichloroethane	0.003	0.005	
Trichloroethylene	zero	0.005	0.05
2,4,6-Trichlorophenol			0.005
Trifluralin			0.045 (IMAC)
Vinyl chloride	zero	0.002	0.002
Xylenes (total)	10	10	
RADIONUCLIDES			
Alpha particles	none	15 picocuries/L	
Americium-241			0.2 Bq/L
Antimony-122			50 Bq/L
Antimony-124			40 Bq/L

	United States		Canada
Contaminant	MCLG (mg/L)	MCL or TT (mg/L)	MAC (mg/L)
Antimony-125			100 Bq/L
Barium-140			40 Bq/L
Beryllium-7			4000 Bq/L
Beta particles and photon emitters	none	4 millirems/year	
Bismurh-210			70 Bq/L
Bromine-82			300 Bq/L
Calcium-45			200 Bq/L
Calcium-47			60 Bq/L
Carbon-14			200 Bq/L
Cerium-141			100 Bq/L
Cerium-144			20 Bq/L
Cesium-131			2000 Bq/L
Cesium-134			7 Bq/L
Cesium-136			50 Bq/L
Cesium-137			10 Bq/L
Chromium-51			3000 Bq/L
Cobalt-57			40 Bq/L
Cobalt-58			20 Bq/L
Cobalt-60			2 Bq/L
Gallium-67			500 Bq/L
Gold-198			90 Bq/L
Indium-111			400 Bq/L
Iodine-125			10 Bq/L
Iodine-129			1 Bq/L
Iodine-131			6 Bq/L
Iron-55			300 Bq/L
Iron-59			40 Bq/L

APPENDIX A: *US and Canadian Standards for Drinking Water* 143

	United States		Canada
Contaminant	MCLG (mg/L)	MCL or TT (mg/L)	MAC (mg/L)
Lead-210			0.1 Bq/L
Manganese-54			200 Bq/L
Mercury-197			400 Bq/L
Mercury-203			80 Bq/L
Molybdenum-99			70 Bq/L
Neptunium-239			100 Bq/L
Niobium-95			200 Bq/L
Phosphorus-23			50 Bq/L
Plutonium-238			0.3 Bq/L
Plutonium-239			0.2 Bq/L
Plutonium-240			0.2 Bq/L
Plutonium-241			10 Bq/L
Polonium-210			0.2 Bq/L
Radium-224			2 Bq/L
Radium-226 and Radium-228 (combined)	none	3 picocuries/L	
Radium-226			0.6 Bq/L
Radium-228			0.5 Bq/L
Rhodium-105			300 Bq/L
Rubidium-81			3000 Bq/L
Rubidium-86			50 Bq/L
Ruthenium-103			100 Bq/L
Ruthenium-106			10 Bq/L
Selenium-75			70 Bq/L
Silver-108m			70 Bq/L
Silver-110m			50 Bq/L
Silver-111			70 Bq/L
Sodium-22			50 Bq/L

Contaminant	United States		Canada
	MCLG (mg/L)	MCL or TT (mg/L)	MAC (mg/L)
Strontium-85			300 Bq/L
Strontium-89			40 Bq/L
Strontium-90			5 Bq/L
Sulphur-35			500 Bq/L
Technetium-99			200 Bq/L
Technetium-99m			7000 Bq/L
Tellurium-129m			40 Bq/L
Tellurium-131m			40 Bq/L
Tellurium-132			40 Bq/L
Thallium-201			2000 Bq/L
Thorium-228			2 Bq/L
Thorium-230			0.4 Bq/L
Thorium-232			0.1 Bq/L
Thorium-234			20 Bq/L
Tritium			7000 Bq/L
Uranium	zero	30 micrograms/L (as of 12/08/03)	0.02 (IMAC)
Uranium-234			4 Bq/L
Uranium-235			4 Bq/L
Uranium-238			4 Bq/L
Ytterbium-169			100 Bq/L
Yttrium-90			30 Bq/L
Yttrium-91			30 Bq/L
Zinc-65			40 Bq/L
Zirconium-95			100 Bq/L

1. Although the MAC for coliforms is zero, because coliforms are not uniformly distributed in water and are subject to considerable variation in public health significance, drinking water conforms to the MAC so long as no sample contains *E. coli*, and no consecutive sample from the same site or not more than ten percent of samples from the distribution system in a given calendar month should show the presence of total coliform bacteria.

Note: One bequerel (Bq) is approximately equal to 27 picocuries (pCi).

Index

2,4D herbicide, 21, 99

A

access to information, 55–57, 67, 89–90
activated alumina treatment, 28
activated carbon adsorption filters, 20, 22, 35, 99, 118–119
activism. *see* consumers
agricultural areas, health risks, 21–22
Alberta, water testing, 54–55, 72
algae, 8, 9
aluminum
   contamination, 16–17
   and water treatment, 16, 33, 118
Alzheimer's disease, 16
anemia, 13, 26
ANSI/NSF standards
   certification, 93
   for contaminants, 98, 99
   for taste and odor reduction, 99
aquifers, 5, 50–51, 79
arsenic, 27–28
Atlanta, private water supplier, 68–69
atrazine, 20–21, 23, 99

B

bacteria, 12, 53, 125
biofilm, 46
   in bottled water, 97
   coliform count, 23, 24, 26
   E. coli, 1, 26–27, 55
   in filtered water, 98, 118
"beaver fever", 25–26
benzene, 96, 99
bisphenol-A, 97
boiled water, 18, 24, 26, 36, 74
boil-water advisories, 36, 74
bottled water, 18, 20, 28
   and baby formula, 96–97
   environmental costs, 97–98
   regulation of, 93–96
   selecting a brand, 24
   summary, 126
   types of, 94

Brita Canada, 98
British Columbia
   disease outbreaks, 22, 74
   water testing, 53

C

C.K. Choi Institute, 83
cadmium, 8, 99
calcium, 8, 9–10, 119
Calgary Waterworks, 63–64
Canada
   complaint procedure, 91
   guidelines, 12–13, 59–61
   outbreak statistics, 74
   provincial standards, 59, 61, 72, 126–127
   testing drinking water, 53–55
   water quality reports, 56–57, 90
   watershed protection, 49, 50–51, 69
cancer, 40
   bladder, 19, 27
   breast, 20, 23
   classification of carcinogens, 14
   colon, 19
   lung, 27, 28
   ovarian, 20
   prostate, 27
   risk factors, 60
   skin, 27
   stomach, 17, 28
carbonation, 8, 96
carbon dioxide, 8, 10
carbon filters. *see* filter systems
carcinogens
   classification of, 14
   and pesticides, 20
   plastic residues, 97
   and trihalomethanes, 19, 91
chemicals. *see also* specific types
   disposal of, 12, 47, 50, 51, 108
   gasoline additive MTBE, 91
   long-term effects, 60–61, 72
children
   and contaminants, 13
   and *E. coli*, 26
   effects of lead, 13
chloramine disinfection, 37

chlorine/chlorination
  by-products, 19ñ20, 36ñ37
  chlorine residues, 8, 9, 36, 93
  and disinfection, 36ñ37
  resistance to, 12, 23, 24, 36
  and trihalomethanes, 19ñ20, 36
chlorine dioxide disinfection, 37
cleaning products disposal, 12, 108
cloudy water. *see* turbidity
coliform count, 23, 24, 26
Colorado River contamination, 29
color of water, 8, 33
Consumer Confidence Reports, 56, 89, 125
Consumer Reports, 82, 84, 97
consumers
  activism, 91ñ92, 127
  complaint procedure, 91
  water quality, 125
contaminants.
  accumulation
  definition, 10
cost of drinking
  examples of,
  global compa
  privatization,
  small water sy
  water metering
cryptosporidiosis
  Cranbrook ou
  Milwaukee ou
  North Battleford outbreak, 25
Cryptosporidium
  contamination, 23ñ24
  inactivating, 37, 38
  outbreak, 25
  reduction, 98

D

desalination, 80
detergents, non-biodegradable, 34
Detroit, private water supplier, 67
diarrheal diseases, 23, 25, 26
dicamba, 21
disease outbreaks
  cryptosporidiosis, 24, 25, 48
  E. coli, 1, 26ñ27, 55
  impact of, 12, 22, 48
  outbreak prevention, 41ñ42
  outbreak statistics, 73ñ74

and sewage, 12
and water distribution, 44
dishwashers, 84
disinfection process. *see also* chlorine
  /chlorination
  chloramines, 37
  chlorine dioxide, 37
  ion exchange, 121, 122
  ozonation, 38ñ39
  requirements of, 35ñ36
  residual effect, 35, 46
  reverse osmosis, 39, 119ñ121
  ultra-violet radiation, 37ñ38, 121ñ122
disposal of hazardous waste, 12, 51, 108
dissolved gases, 8
dissolved minerals, 7ñ8
distillation system, 17, 28
distilled water, 7
distribution systems, 43ñ46. *see also* pipes
drinking water
  backflow problems, 44ñ46
  Canada guidelines, 12ñ13, 59ñ61
  characteristics, 7ñ10
  consumption rates, 77ñ78
  cost of, 41, 61ñ69
  household reduction, 80ñ86
  regulation, 12ñ13
  sample monitoring schedule, 54
  source protection, 5, 47ñ51, 127
  testing, 53ñ55, 122ñ123
  use by sector chart, 78
  US standards, 13, 60ñ61
  water quality reports, 56ñ57, 67, 89ñ90, 125
drugs, pharmaceutical, 10, 12, 23
DTE Energy, 65

E

E. coli
  contamination, 26ñ27
  Walkerton outbreak, 1, 27, 55
Edmonton, private water supplier, 57, 65
effluent
  agricultural, 17, 24, 47, 49ñ50
  and disease, 12, 22ñ23, 25
  in drinking water, 10
  industrial, 29, 47, 50
  runoff, 11, 47ñ51
emergency response
  "boil water order", 36, 74
  planning for, 42

and private water suppliers, 67
environmental issues
  bottled water, 97–98
  bulk water exports, 69–70, 97–98
Environmental Protection Agency (EPA)
  enforcement, 59, 73
  sample monitoring schedule, 54
  standards setting, 60–61
  surface water protection, 49–50
estrogen and estrogen mimics, 23, 97
exports, bulk water, 69–70, 97–98

F

faucet-mounted filters, 99, 118
fertilizers. *see* nitrates
fetal development, 75
  birth defects, 17, 19
  "blue baby syndrome", 17
  low birth weight, 13, 19
  miscarriages, 17, 19, 91
  still births, 13
filter systems
  activated carbon adsorption, 20, 22, 35, 99, 118–119
  ANSI/NSF certification, 93
  comparison chart, 120
  distillation, 17, 28
  faucet-mounted, 99, 118
  jug filters, 98–99
  particulate filters, 118
  point-of-use, 99
  reverse osmosis, 17, 18, 24, 28, 39, 119–121
  showerheads, low-flow, 84
  summary, 126
filtration systems
  rapid gravity, 34–35
  slow sand, 33–34, 116–117
First Nation water systems, 74
fluoride/fluoridation, 8, 39–41, 119
Food and Drug Administration (FDA)
  bottled water regulations, 94–95
freshwater
  bulk water trade, 69–70
  sources, 3–4

G

gardens, watering tips, 85–87
Giardia
  contamination, 25–26
  inactivating, 36, 37, 38

Giornelli, Greg, 69
greywater reuse, 86–87
groundwater, 5
  collection systems, 101–108
  contamination, 11, 19, 21
  and pH, 10
  purification process, 31
  saltwater intrusion, 79–80
  and watershed protection, 47–51

H

haloacetic acids (HAAs), 36
"hamburger disease", 26–27
Hamilton, private water supplier, 67–68
hard water, 9–10, 121, 122
Health Canada
  bottled water regulations, 93–94
  drinking water quality guidelines, 59, 60–61
health risks. *see also* disease outbreaks
  children, 13, 26
  hot tap water, 15
  infants, 17, 19, 74–75, 96–97
  long-term, 1, 60–61, 72
  low immunity, 24, 25, 74, 125
  and pipes, 13, 44, 45–46
  public awareness, 1, 125–127
heart disease, 10
hormones and hormone mimics, 23
household chemicals disposal, 12, 51, 108
household plumbing. *see* pipes
hydrogen sulfide, 9
hydrological cycle, 4–5, 11–12
hypertension, 13

I

immune system
  compromised, 24, 25, 74, 125
  of infants and children, 13, 74
infants
  baby formula, 17, 96–97
  and boiled water, 74
  effects of lead, 13, 74–75
  effects of nitrates, 17, 74–75
  effects of trihalomethanes, 19
infrastructure, cost of, 41, 46, 62, 67
ion exchange filtration
  disinfection process, 121, 122
  effectiveness, 18, 28, 47
iron, 8, 9, 118

J
jug filters, 98ñ99

K
kidney dysfunction, 13, 26, 27
kidney tumors, 27

L
land use, 48ñ51
lawns, water-efficient, 85ñ86
lawsuits, 73, 91
lead, 8, 99
   checking for lead pipework, 15
   contamination, 13ñ16, 75, 125
   leaching, 13, 44
   and soft water, 10
Lee County (FL), private water supplier, 68
logging, watershed, 49

M
magnesium, 8, 9ñ10
manganese, 8, 9, 118
Maximum Acceptable Concentrations (MACs), 12ñ13, 61
Maximum Contaminant Level Goal (MCGL), 60, 90
Maximum Contaminant Levels (MCLs), 13, 59, 60, 90
membrane filters. *see* reverse osmosis filters
mercury, 8, 99
mineral water, 7
miscarriages, 17, 19, 91
Moncton, private water supplier, 66
monitoring
   for contaminants, 53ñ55
   of regulation violations, 73
   water pressure, 46
municipal water suppliers
   cost comparison, 61ñ62
   privatization, 64ñ69
   small communities, 41, 66, 72ñ74, 125
   violations, 73ñ74
   water metering, 62ñ64

N
New Brunswick, watershed, 49, 51
Newfoundland
   disease outbreaks, 74
   watershed protection, 51

New York City, watershed, 49ñ50
nitrates, 8
   contamination, 17ñ18, 75
   removal, 47, 119, 121
non-Hodgkin's lymphoma, 17
North American Free Trade Agreement (NAFTA), 69ñ70
Nova Group, 69

O
O'Connor, Dennis, 1
odor of water, 9
   disinfection process, 37, 38
   and filter systems, 99
organic material
   effect on pH, 10
   effect on water appearance, 8
   screening, 32
   and trihalomethanes, 19
organic waste
   source of *E. coli*, 26, 27
   source of nitrates, 17
   source of parasites, 24, 25
   source of pathogens, 22ñ23
osteoporosis, 40
ozonation
   disinfection process, 38ñ39
   effectiveness, 20, 22, 24, 26

P
paints and solvents disposal, 12, 51, 108
parasites, 23ñ26, 36, 37, 38, 53
pathogens, 125
   bacteria, 12, 53, 125
   boil-water advisories, 36, 74
   chlorine-resistant, 12, 23, 24, 36
   coliform count, 23, 24, 26
   Cryptosporidium, 23ñ24, 25, 37, 38
   E. coli, 1, 26ñ27, 55
   in filtered water, 98, 118
   Giardia, 25ñ26, 36, 37, 38
   types of, 22ñ23
   viruses, 12, 53, 125
perchlorate, 29, 60
Perrier bottled water, 96
pesticides
   contamination, 20ñ22
   hormone mimics, 23
   removal, 22, 38, 118, 119
   seasonal peaks, 21ñ22, 71ñ72
pH, 10, 13, 14, 38

Index 149

pipes
  corrosion inhibitors, 39
  corrosion of, 10, 44
  cross-connections and backflow, 44ñ46
  distribution system, 43ñ46
  hard water deposits, 9
  health risks, 45ñ46
  lead leaching, 13, 44
  lead pipe identification, 15
plastic residues, 97
plumbing. *see* pipes
point-of-use filters, 99
pollution, water
  effluent, 10ñ11, 29, 47, 50
  household chemicals, 12, 51, 108
potassium, 96
pregnancy. *see* fetal development
private water suppliers, 62, 64ñ69
private water systems
  disinfection process, 115ñ122
  planning for, 101ñ102
  rainwater collection, 110ñ114
  springs, 108, 109
  storage, 114ñ115
  surface water collection, 108ñ110
  testing drinking water, 122ñ123
  wells, 103ñ108
privatization
  corporate pressure, 64ñ65
  effect of, 66ñ69
  public-private partnerships, 65ñ66
public awareness. *see* consumers
public-private partnerships, 65ñ66
public water suppliers. *see* municipal water suppliers

R

radon, 28ñ29, 60, 99
rainwater collection, 86, 110ñ114
reverse osmosis filters
  disinfection process, 39, 119ñ121
  effectiveness, 17, 18, 24, 28

S

Safe Drinking Water Act (US), 29, 50, 56, 59, 126
Seattle, UV radiation system, 38, 65
septic systems, 108
shortages, water, 4, 79ñ80
showerheads, low-flow, 84

small community water suppliers
  cost of, 41, 66
  health risks, 72, 125
  violation rate, 73ñ74
sodium, 119
softeners, 121, 122
soft water, 10, 13, 14
springs, 108, 109
ST Environmental Services, 68
storage, water, 43ñ44, 114ñ115
strontium, 9
sulfates, 9, 96
Sun Belt Water, 69
surface water, 5
  collection systems, 108ñ110
  contamination, 19, 22, 24, 108
  purification process, 31
  treatment process, 32
  and watershed protection, 48ñ51
Surface Water Treatment Rule (US), 49

T

tap water. *see* drinking water
taste of water
  causes of, 8, 9, 10
  chlorine dioxide disinfection, 37
  and chlorine residues, 9, 36, 93
  and filter systems, 99
  minerals, 7ñ8
  ozonation, 38
testing drinking water
  home test kits, 90
  laboratories, 55, 123
  private water systems, 122ñ123
THMs. *see* trihalomethanes (THMs)
thyroid tumors, 29
toilets
  compost, 82ñ83
  leaks, 85
  low-flow, 81ñ82
  toilet dams, 83
Toronto Healthy House, 111ñ112
trade, bulk water, 69ñ70
trihalomethanes (THMs)
  and Chesapeake lawsuits, 91
  contamination, 19ñ20, 36ñ37
turbidity, 8, 33, 118

U

ultra-violet radiation

disinfection process, 37–38, 121–122
effectiveness, 20, 24, 26
United States
complaint procedure, 91
lead in water distribution system, 13
outbreak statistics, 73–74
standards, 13, 60–61
testing requirements, 53–55
water quality reports, 55–56, 67, 89–90, 125
United Water, 68

V

Vancouver (BC)
water quality reports, 57
and watershed protection, 49
Victoria (BC), 38
water quality reports, 57
and watershed protection, 49
viruses, 12, 53, 125

W

washing machines, 84
wastewater reuse, 86–87
water conservation
consumption rates, 77–78
rainwater collection, 86, 110–114
reusing wastewater, 86–87
shortages, 4, 79–80
toilets, high-efficiency, 80–83
water metering, 62–64
water saving tips, 84–86
water coolers, 97
water cycle, 4–5, 11–12
watering, outdoors, 85–87
water metering, 62–64
water pressure, 44–46, 114
water quality reports, 55–57, 67, 89–90, 125
watershed protection, 48–51, 69
water shortages, 4, 79–80
water softeners, 121, 122
water treatment systems. *see also* disinfection process
cost of, 41
disinfection process, 35–39
filtration systems, 33–35, 116–117
fluoridation, 39–41
natural processes, 31
preliminary stage, 32–33
water use. *see* drinking water
wells, abandoned, 50, 108
wells, private
digging, 106–107
installation, 102
maintenance, 107–108
types of, 103–106
well water
contamination, 17, 22, 91
"good neighbor" systems, 41
protection of, 50
World Health Organization (WHO)
classification of carcinogens, 14

About the Author

JULIE STAUFFER is a freelance writer and editor in Guelph, Ontario specializing in health, science and environmental issues. She has a Master's degree in biology from the University of Waterloo and a certificate in magazine journalism from Ryerson Polytechnic University in Toronto.

She has worked in the environmental sector for a number of years, including stints as an associate editor at *Alternatives Journal* and communications coordinator for the Sierra Club of Canada, BC Chapter.

Julie is the writer and co-producer of *The Big Flush*, an educational video on sewage issues, and she has published two books on water issues: *The Water Crisis: Constructing solutions to freshwater pollution*, and *Safe to Drink?*, a guide for UK consumers on the quality of their tap water.

If you have enjoyed *The Water You Drink*,
you might also enjoy other

BOOKS TO BUILD A NEW SOCIETY

Our books provide positive solutions for people who want to make a difference. We specialize in:

Natural Building & Appropriate Technology

Sustainable Living • Ecological Design and Planning

Environment and Justice • Conscientious Commerce • New Forestry

Educational and Parenting Resources • Nonviolence

Progressive Leadership • Resistance and Community

New Society Publishers

ENVIRONMENTAL BENEFITS STATEMENT

New Society Publishers has chosen to produce this book on Enviro 100, recycled paper made with **100% post consumer waste**, processed chlorine free, and old growth free.

For every 5,000 books printed, New Society saves the following resources:[1]

21	Trees
1,902	Pounds of Solid Waste
2,093	Gallons of Water
2,730	Kilowatt Hours of Electricity
3,458	Pounds of Greenhouse Gases
15	Pounds of HAPs, VOCs, and AOX Combined
5	Cubic Yards of Landfill Space

[1] Environmental benefits are calculated based on research done by the Environmental Defense Fund and other members of the Paper Task Force who study the environmental impacts of the paper industry.

For more information on this environmental benefits statement, or to inquire about environmentally friendly papers, please contact New Leaf Paper – info@newleafpaper.com Tel: 888 • 989 • 5323.

For a full list of NSP's titles, please call 1-800-567-6772 *or check out our web site at:*
www.newsociety.com

NEW SOCIETY PUBLISHERS